THE ENDOMETRIUM
Hormonal Impacts

THE ENDOMETRIUM
Hormonal Impacts

EDITED BY

JEAN de BRUX
Foundation for Education and Research in Histo- and Cytopathology
Paris, France

RODRIGUE MORTEL
M. S. Hershey Medical Center, Pennsylvania State University
Hershey, Pennsylvania

JEAN PIERRE GAUTRAY
University of Paris Val de Marne
Creteil, France

SPRINGER SCIENCE+BUSINESS MEDIA, LLC

Library of Congress Cataloging in Publication Data

Main entry under title:

The Endometrium, hormonal impacts.

"Proceedings of a symposium on the physiology and pathophysiology of the menstrual cycle,
held April 22-23, 1980, in Paris, France"—Verso t.p.
  Bibliography: p.
  Includes index.
  1. Menstrual cycle—Congresses. 2. Hormones, Sex—Physiological effect—Congresses. 3.
Endometrium—Congresses. I. Brux, Jean de. II. Mortel, Rodrigue. III. Gautray, Jean Pierre.
|DNLM: 1. Endometrium— Physiology—Congresses. 2. Endometrium—Physiopathology—
Congresses. 3. Menstruation—Congresses. 4. Sex hormones—Physiology—Congresses. WP
400 E56 1980|
QP263.E5   612'.662   81-8529
                                                                                        AACR2

Proceedings of a symposium on the physiology and pathophysiology of the
menstrual cycle, held April 22–23, 1980, in Paris, France

ISBN 978-1-4684-3979-3   ISBN 978-1-4684-3977-9 (eBook)
DOI 10.1007/978-1-4684-3977-9

© 1981 Springer Science+Business Media New York
Originally published by Plenum Press, New York in 1981
Softcover reprint of the hardcover 1st edition 1981

PREFACE

    Although the physiology of the menstrual cycle appears clear
and easily explained by a balance in the concentration of various
sex steroid hormones, numerous details of its mechanism are still
poorly understood and little is known about the relationship
among clinical events, plasma hormone concentrations, molecular
impacts on target tissues and their regulation. In the following
chapters, the authors have attempted to establish a correlation
between endometrial histology and well-understood physiologic
events of the menstrual cycle. They have provided up-to-date
information on the effects of various hormones and combinations
of hormones on the endometrium. The interdependence of endometrial
morphology, molecular biology, endocrinology and physiology,
provides grounds for a better understanding of the complex mechanism
of the menstrual cycle, and sheds some light on its pathophysiology.
Such an approach adds another dimension to interpretation of many
menstrual abnormalities and numerous aspects of infertility in
women with normal physiognomies and apparently regular menstrual
cycles.

    The pathologist must be aware of these new concepts since
a knowledge of functional changes reflected in hormone serum
levels and sex steroid receptor concentrations allows a more
detailed analysis and a better interpretation of the structural
features of the endometrium. This information placed in the
proper clinical context can help the gynecologist provide optimal
therapy. The reader will find a valuable reference for a synthesis
of clinical, morphological, and biochemical data related to the
menstrual cycle and its aberrations.

    We wish to thank the Roussel and Carlo Erba laboratories
for their support in the organization of the Symposium. We
acknowledge the excellent cooperation of the contributors and
the valuable effort of Carol Kanofsky in the preparation of the
manuscript. Finally, we appreciate the editorial assistance of
Melissa Reese.

Jean de Brux
Rodrigue Mortel
Jean Pierre Gautray

CONTENTS

MENSTRUAL CYCLE: SOME UNCERTAIN ASPECTS

Jean Pierre Gautray

Department of Obstetrics and Gynecology
Université de Paris Val de Marne
C.H.I.C., 94010 Creteil Cedex
France

INTRODUCTION

For a long time endometrial biopsy was the only unbiased and
precise parameter for clinical investigation of the human menstrual
cycle. With the development of techniques to measure hormone concentrations in plasma, the earlier methodology was abandoned.
Interest in this target tissue is again growing, but for different
reasons. Because hormonal measurements indicate a momentary physiological value, whereas the endometrium reflects various stages
of hormonal influence, sampling and histological examination of this
tissue can demonstrate the normal correlated influences of hypophysial and ovarian hormones, or show a functional disorder of the
hypothalamic-ovarian axis. New insights into the cellular and
molecular physiology of the endometrium are probably necessary for
understanding implantation mechanisms. This paper will be devoted
to the effects of hormones on the endometrium. Rather than a
sempiternal report on the known physiological aspects of the menstrual cycle, we will focus on the unknown aspects of synchronized
events that control human reproduction. It does not seem useful to
describe the concentrations of the different hypophysial and ovarian
hormones or necessary to relate the feedback system or the hormonal
mode of action at the cellular level. Instead we will attempt to
describe the controversial aspects of menstrual cycle regulation,
and to examine specific and still unclear points of this regulation.

CENTRAL LEVEL

Hypophysial Activity and Sensitivity

Control of the menstrual cycle is much more complex than the classic and well demonstrated estrogen-induced feedback mechanism (Bogdanove, 1963; Knobil, 1974). Two phenomena are most important for control of the cycle in primates. The first is the variable hypophysial sensitivity to luteinizing hormone releasing hormone (LHRH) during the cycle; administration of repetitive doses of LHRH (10 µg at 2-hour intervals, 5 successive times) during the menstrual cycle produces profound changes in pituitary response, in concert with cyclic ovarian steroid levels (Wang et al., 1976). During the late follicular phase, subsequent responses to successive LHRH doses are greater. The LH response is maximal during the preovulatory period, whereas gonadotropin rises show a progressive decrease after injection of LHRH. The LH response then decreases during the luteal phase (Yen and Lein, 1976). These variations in hypophysis responsiveness are clearly dependent on circulating estrogen (Yen and Tsai, 1972; Jaffe and Keye, 1974; Young and Jaffe, 1976). The second phenomenon is the autonomy of the mediobasal hypothalamus, which is not dependent on a signal from the preoptic area for induction of the preovulatory LH peak (Krey et al., 1975). Thus, it is not surprising that no LHRH surge could be detected before or during the preovulatory LH peak (Barraclough, 1979). With this experimental and clinical background, we wanted to examine how the preovulatory LH peak is determined in primates.

Yen et al. (1972) have suggested that there are two pools of pituitary gonadotropins. One would be readily releasable and discharged in response to a brief small pulse of LHRH. Indeed, changing pituitary responses to pulses of different amounts of LHRH (10 to 300 µg) demonstrate that the pituitary can detect relatively small variations in circulating LHRH. A second reserve pool would then be activated and released after a more prolonged LHRH stimulation. This pool would be dependent on the level and duration of estradiol secretion (Yen et al., 1972; Wang et al., 1976; Hoff et al., 1977).

Based on these studies, a hypothesis integrating the total regulation of the menstrual cycle can be postulated. The sensitivity of the hypophysis would be enhanced by the rhythm of estradiol secretion, i.e., the higher the estradiol level, the more important gonadotropin release. Hypothalamic LHRH secretion would be essential in that it would have a permissive, but obligatory role (Yen, 1978). In this context the ovary would be the real timer regulating this complex system. This hypothesis is perhaps oversimplified; however, recently the essential permissive role of LHRH pulses has been demonstrated (Knobil, 1980a,b). Female rhesus monkeys were subjected to bilateral irradiation, producing lesions in the arcuate region of the mediobasal hypothalamus and suppressing gonadotropin-releasing hormone (GnRH) production. This resulted in the reduction of LH, FSH and estrogen and in cessation of the menstrual cycle. Hormonal secretions and the menstrual cycle resumed upon administration of pulses of GnRH on an hourly regimen (Knobil et al., 1980). A similar GnRH infusion regimen induced normal ovulatory-menstrual cycles in prepubertal female rhesus monkeys; when infusions were discontinued, the animals promptly reverted to their prepubertal state (Wildt et al., 1980).

These experimental data have had few but interesting clinical applications (Crowley and McArthur, 1980; Hanker and Schneider, 1980; Layendecker et al., 1980). A few cases of hypothalamic amenorrhea have been at least temporarily corrected by hourly pulsatile LHRH administration, and two pregnancies resulted.

Previous attempts to modify hypophysis sensitivity or to stimulate gonadotropin secretion have been disappointing. Administration of estradiol or synthetic estrogen (ethinyloestradiol, moxestrol) during the follicular phase usually delays ovulation and subsequently shortens the luteal phase (Carguille et al., 1973; Gautray et al., 1977). Administation of LHRH and potent analogs (at various doses) has not improved ovulatory disorders. LHRH pulsatile infusion, although difficult, may be very promising clinically. It demonstrates the permissive role of LHRH secretion; however, the origin of this hypothalamic activity remains unknown.

Role of Prolactin

The role of prolactin (PRL) in primate reproduction is not well understood. Variations in plasma PRL levels during the menstrual cycle are controversial (McNeilly and Chard, 1976; Gautray et al., 1977; Vekemans et al., 1977); however there is no evidence of a consistent change in PRL during the menstrual cycle. Whether PRL is necessary for maturation of the ovarian follicle is unknown, although hyperprolactinemia, either "dysfunctional" or due to pituitary adenoma (Boyd and Richlin, 1978) can induce (or be associated with) different menstrual cycle disorders from amenorrhea to luteal insufficiency (Corenblum et al., 1976; Seppala et al., 1976; Muhlenstadt et al., 1978; Del Pozo et al., 1979). The mechanism of hyperprolactinemia without clinically demonstrable pituitary adenoma is difficult to ascertain, but may be of hypothalamic origin or related to a specific disorder of the PRL-producing cells. There is no means to distinguish between neural or cellular disorders, and to date, pharmacological tests have been unreliable (Genazzani et al., 1980).

Role of Neurotransmitters

Results of studies on rodents support the hypothesis that amines and neuropeptides are involved in the control of pituitary hormone secretion (Boyd and Reichlin, 1978; Fuxe et al., 1978; Meites et al., 1979). The best known brain modulation (by mono-aminergic mechanisms) concerns dopamine (DA) and PRL (Boyd and Reichlin, 1978). Morphological studies have demonstrated that DA, norepinephrine (NE) and LHRH receptors are concentrated in the median eminence, where these substances are present in high concentrations. This supports the hypothesis of a possible interaction between DA, NE and LHRH receptors and a possible monoaminergic control of gonadotropin secretion.

In the rat, a decreased DA turnover, a sharp increase in NE turnover during the afternoon of proestrus (Ben-Jonathan et al., 1977) and a significant increase in NE concentration between the morning and afternoon of proestrus in hypothalamic nuclei (Selmanoff et al., 1977) have been demonstrated. These data suggest that DA has an inhibitory effect, whereas NE has an enhancing effect on the control of LHRH secretion. In humans, these mechanisms are much more difficult to examine. Most enzyme inhibitors of catecholamine metabolism are toxic and cannot be used. DA infusions (4 µg/kg/min) have been used because of DA receptors on pituitary cells, the extra meningeal location of the gland (Leblanc et al., 1976; Lachelin et al., 1977; Judd et al., 1978) and because they significantly lower PRL and LH levels. PRL variations appear to be dependent on patient category (male, female, hyperprolactinemic), and on the day of menstrual cycle (Leblanc et al., 1976; Judd et al., 1978). LH variation is dependent on the cycle day and consequently on estrogen (Leblanc et al., 1976; Judd et al., 1978); however, FSH concentrations are less sensitive to this hormonal influence (Leblanc et al., 1976; Judd et al., 1978). Although these data demonstrate pituitary sensitivity to DA, it is difficult to demonstrate the inhibitory role of DA on gonadotropins or LH secretion. Other clinical data do not allow simple and precise conclusions. Bromocriptine, a potent DA agonist, induces a very sharp decline in PRL concentration, but does not inhibit LH release in hyperprolactinemic women (Evans et al., 1980), and has been successfully used in prolactinemic patients for treatment of anovulation and luteal insufficiency (Peillon et al., 1979). In patients who recently reached menopause, we performed a pituitary stimulation test (simultaneous 250 µg thyroid-releasing hormone and 200 µg LHRH bolus) after administration of bromocriptine or haloperidol (a DA antagonist). No variation in LH was osbserved, although PRL was inhibited by bromocriptine and stimulated by haloperidol (unpublished data). These contradictory results support the hypothesis that DA and NE modulate rather than directly influence LH and PRL levels, at least in humans. More precise pharmacologic investigations are necessary.

Present knowledge concerning other neurotransmitters and pharmacologic agents is not clear. Gamma-aminobutyric acid (GABA) and its metabolite gamma-hydroxybutyric acid (GHB) may affect gonadotropin, growth hormone and PRL secretion (Ondo, 1974; Takahara et al., 1977). Serotonin is probably involved in stimulation of PRL secretion (Boyd and Reichlin, 1978; Nathan et al., 1980), and some other pituitary hormones (Modlinger et al., 1980). The role of endogenous opiates studied both in vivo and in vitro by physiological and pharmacological methods (Bruni et al., 1977; Shaar et al., 1977; Morley et al., 1980; Quigley and Yen, 1980; Van Loon et al., 1980) indicates that they are probably modulators of these pituitary hormone secretions. However, inhibition or stimulation of these substances may be more widespread than simply affecting a group of specific neurons.

OVARIAN LEVEL

Different physiological aspects of ovarian activity are still puzzling. Three of these, follicle development, ovarian regulation of the reproductive process, and corpus luteum evolution and life-span, will be discussed.

Follicular Growth and Development

Because of investigations on the control of follicular growth and development, the general process of follicular maturation is generally known. Briefly, LH and FSH stimulate follicle growth, estrogen secretion by the theca, and granulosa cell growth. In the beginning, granulosa cells divide and acquire FSH receptors, and then FSH stimulates the appearance of LH receptors. By the time of the preovulatory LH surge, the granulosa cells have enhanced ability to luteinize if removed from the follicle and cultured (Channing et al., 1978; Bjersing, 1979). Nevertheless, why one follicle grows, becomes dominant, and ovulates during each menstrual cycle remains essentially unknown. In primates and humans, only one follicle matures and ovulates; most undergo atresia, even though all follicles are exposed to the same blood level of LH and FSH. Whether the control mechanism of follicular evolution resides within the ovary itself or is dependent on gonadotropin secretion is unknown.

Several investigations in monkeys have examined the influence of gonadotropin on follicular regulation, since the hormonal processes are similar in monkeys and humans. Human menopausal gonadotropin (hMG), administered during the early or midfollicular phase, increases the number of follicles stimulated. Later in the follicular phase after a dominant follicle has developed, there is no further supplementary follicle development, and after destruction of the dominant follicle, the other follicles are temporarily less responsive to hMG (di Zerega and Hodgen, 1980a). Under experimental conditions, selection of a dominant follicle is not altered by low FSH levels, although an increase in FSH is observed in normal cycles at the end of corpus luteum activity (di Zerega et al., 1980).

In women, the number of atretic follicles increases during the luteal phase and is considered progesterone dependent, due to quantitative and/or qualitative changes in gonadotropin secretion, or to a modified ovarian responsiveness. However, this apparent ovarian refractoriness to gonadotropin can be easily overcome (di Zerega and Hodgen, 1980b), at least in monkeys. Investigations suggest that intraovarian control of follicular function is probably important. Intraovarian sex steroids produced by the follicle itself, have been shown to play a role in this regulation in rodents, i.e. exogenous estrogen stimulates growth and prevents follicle atresia, whereas androgen inhibits these effects and promotes atresia (Louvet et al., 1975). The high levels of aromatase activity in granulosa cells and estradiol concentration in pre-ovulatory follicles contrast with their low levels in immature and/or atretic follicles. Aromatase activity is FSH dependent, and in vitro is inhibited by naturally occurring androgen metabolites that have been identified in the human ovary (5-alpha-androstane-3, 17-dione; Batta et al., 1980; Hillier et al., 1980). The role of the aromatase enzyme system is supported by these investigations and participation of 5-alpha-reductase enzyme activity is suspected.

A few nonsteroidal, polypeptide, intraovarian messengers have been detected, although their role and importance have still not been determined. An oocyte maturation inhibitor able to inhibit the spontaneous resumption of meiosis, and a luteinization inhibitor have been extracted from porcine follicles (Channing, 1979). Cow and pig follicular fluids have a potent FSH inhibitory action, and castration of rats is immediately followed by a rise in FSH that cannot be suppressed by estrogen. This potent anti-FSH substance, which is also detectable in rete testis fluid, is called inhibin. Its polypeptide structure is not yet clearly defined.

Follicle growth regulation requires both pituitary and intra-ovarian messengers. Plasma levels of pituitary hormones are well known, and insufficient FSH secretion has been demonstrated in menstrual cycle disorders, especially luteal insufficiency. The role and mechanism of intraovarian factors are not sufficiently known to be used in clinical practice, although they could become contraceptive agents.

Corpus Luteum: Luteinization and Luteolysis

     Immediately after ovulation, luteinization breaks down the
corpus luteum. Its constitutive process, lifespan and progressive
loss of activity are still incompletely understood. Luteinization
involves two different aspects of granulosa cell activity; one is
biochemical (high progesterone secretion) the other is morphological
(transformation) (Westwerdt et al., 1979). Many recent investigations have demonstrated that granulosa cell multiplication and
evolution begin early during follicular maturation (Dimino et al.,
1979) and are inhibited by follicular fluid. These cells then
burst immediately after ovulation. Such a correlation in time
indicates that the corpus luteum quality and lifespan are dependent on follicular growth (Vande Wiele and Turksoy, 1965). It is
presently thought that corpus luteum activity is dependent on
luteotropic factors (Niswender et al., 1972; Ross and Hillier,
1978). Among these, the role of PRL is controversial (McNatty
1979a,b), but that of LH is essential. From the time of LH peak,
progesterone is preferentially secreted compared with 17-OH-preg-
nenolone, multiplication of granulosa cells stops and their transformation into luteinized cells is enhanced (Thibault and Levasseur,
1979). Accessory luteotropic factors include estradiol (at least
in rats) and prostaglandins of the E type (Thibault and Levasseur,
1979).

     The corpus luteum lifespan is 14 ± 2 days in women. Its
regression begins a few days earlier and is marked by a progressive
or precipitous decline in progesterone secretion. "Regression of
the corpus luteum is pivotal in regulating the estrous and menstrual
cycle since it permits follicular development and the ovulatory
surge of pituitary gonadotropins in the subsequent cycle." (Behrman
et al., 1979).

     The first demonstration of luteolysis was the influence of
hysterectomy and endometrium-secreted prostaglandin $F_2\alpha$ in the rat,
sheep, and horse. After hysterectomy, corpus luteum activity is
maintained; endometrial ablation has the same effect due to the
anatomical position (ipsilateral) of the uterine vein and the
ovarian artery (Thibault and Levasseur, 1979). The influence of
endometrium-secreted prostaglandin is demonstrated by its high
concentration in the uterine vein and the opposite effect of indomethacin upon injection into the uterine horn (Thibault and Levasseur, 1979). Recently, different cellular aspects of this luteolytic
mechanism have been set forth: prostaglandin $F_2\alpha$ impedes gonadotropin uptake in vivo and in vitro, the adenylate complex is no
longer observed and LH binding is reduced (Behrman et al., 1979).

     In primates, the cellular mechanism is probably similar, since
prostaglandin receptors have been demonstrated (Powell et al.,
1974); however, the physiological mechanism is different, as endo-

metrial prostaglandins (Singh et al., 1975) cannot reach the ovary
directly, and it is estradiol dependent (Auletta et al., 1978).
In cattle and probably primates, progesterone, estradiol and prostaglandin are produced by the same cell (Shemesh and Hansel, 1975).
Nevertheless, in primates prostaglandin $F_{2\alpha}$ and indomethacin produce no physiological or clinical effect.

Another cellular mechanism that contributes to luteolysis
regulation is cell desensitization or down-regulation, which is
concerned with regulation of peptide hormone receptors and target
cell responses due to increased concentrations of homologous hormone (Catt et al., 1979). This process assists in the regulation
of receptor sites in target cells by polypeptide hormones and is
particularly marked in the testis and ovary after exposure to
elevated gonadotropin levels. In the rat ovary, it follows
several successive phases (Catt et al., 1979). The early phase of
receptor site occupation is characterized by acute biological
stimulatory responses (1 min to 3 hr), activation of adenylate
cyclase, generation of cyclic AMP, and production of progesterone.
During the late phase of site occupation, the primary refractory
state begins (6 hr), the number of free sites, adenylate cyclase
responsiveness, cyclic AMP response, and progesterone production
decrease. The secondary refractory state (1 to 4 days) is characterized by the loss of LH receptors, loss of adenylate cyclase
stimulation, and loss of cyclic AMP response and progesterone
secretion. A recovery period follows this refractory state and
lasts 3 to 10 days (Catt et al., 1979). LH receptors are probably
effective only once, and then are degraded by intracell resorption
and "internalization" (Catt et al., 1979; Lindner, 1979).

In summary, continued exposure of ovarian cells to gonadotropins or prostaglandin results in a characteristic sequence of
events: (1) increased cyclic AMP formation; (2) progressive
desensitization of the adenylate cyclase system, attended or
followed by down-regulation of receptor density on the cell surface;
and (3) gradual recovery of responsiveness upon removal of the
hormone (Lindner, 1979).

This membrane regulatory mechanism appears to be of great
importance in luteolysis and could also participate in corpus
luteum maintenance. Pulsatile LH secretion would be significant
as each LH pulse would determine the occupation of a few binding
sites, their clustering, and their degradation by "internalization",
and the loss of a few receptor sites. The disappearance of successive sites would stimulate continuous receptor synthesis. If
this hypothesis is correct, repeated tiny desensitizations would
have great importance in corpus luteum activity (Lindner, 1979).

REFERENCES

Auletta, F.J., Agins, H., and Scommegna, A., 1978, Prostaglandin F mediation of the inhibitory effect of estrogen on the corpus luteum of the rhesus monkey, Endocrinology, 103:1183.

Barraclough, C.A., 1979, Central nervous regulation of the pre-ovulatory release of FSH and LH from the pituitary gland, in: "Human Ovulation," E.S.E. Hafez, ed., North Holland Publishing Co., Amsterdam, p. 101.

Batta, S.K., Wentz, A.C., and Channing, C.P., 1980, Steroidogenesis by human ovarian cell types in culture: influence of mixing of cell types and effect of added testosterone, J. Clin. Endocrinol. Metab., 50:274.

Behrman, H.R., Luborsky-Moore, J.L., Pang, C.Y., Wright, K., and Dorflinger, L.J., 1979, Mechanisms of $PGF_{2\alpha}$ action in functional luteolysis, Adv. Exp. Med. Biol., 112:557.

Ben-Jonathan, N., Oliver, C., Weiner, H.J., Mical, R.S., and Porter, J.C., 1977, Dopamine in hypophysial portal plasma of the rat during the estrous cycle and throughout pregnancy, Endocrinology, 100:452.

Bjersing, L., 1979, Intraovarian mechanisms of ovulation, in: "Human Ovulation," E.S.E. Hafez, ed., North Holland Publishing Co., Amsterdam, p. 149.

Bogdanove, E.M., 1963, Direct gonad-pituitary feedback: an analysis of effects of intracranial estrogenic depots on gonadotropin secretion, Endocrinology, 73:696.

Boyd, A.E., and Reichlin, S., 1978, Neural control of prolactin secretion in man, Psychoneuroendocrinology, 3:113.

Bruni, J.F., Van Vugt, D., Marshall, S., and Meites, J., 1977, Effects of naloxone, morphine and methionine-enkephalin on serum prolactin, luteinizing hormone, follicle stimulating hormone, thyroid stimulating hormone and growth hormone, Life Sci., 21:461.

Carguille, C.M., Vaitukaitis, J.L., Bermudez, J.A., and Ross, G.T., 1973, A differential effect of ethinyloestradiol upon plasma FSH and LH relating to time of administration in the menstrual cycle, J. Clin. Endocrinol., 36:87, 1973.

Catt, K.J., Harwood, J.P., Richert, N.D., Conn, P.M., Conti, M., and Dufau, M.L., 1979, Luteal desensitization: hormonal regulation of LH receptors, adenylate cyclase and steroidogenic responses in the luteal cell, Adv. Exp. Med. Biol., 112:647.

Channing, C.P., 1979, Follicular nonsteroidal regulators, Adv. Exp. Med. Biol., 112:327.

Channing, C.P., Anderson, L.D., and Batta, S.T., 1978, Follicular growth and development, Clin. Obstet. Gynecol., 5:375.

Corenblum, B., Pairaudeau, N., and Shewshuk, A.B., 1976, Prolactin hypersecretion and short luteal defects, Obstet. Gynecol., 4:486.

Crowley, W.F., and McArthur, J.W., 1980, Stimulation of the normal menstrual cycle in Kallman's syndrome by pulsatile administration of LHRH, J. Clin. Endocrinol. Metab., 5:173.

Del Pozo, E., Wyss, H., Tolis, G., Alcaniz, J., Campana, A., and Naftolin, F., 1979, Prolactin and deficient luteal function, Obstet. Gynecol., 53:282.

Dimino, M.J., Elfont, E.A., and Berman, S.K., 1979, Changes in ovarian mitochondria: early indicators of follicular luteinization, Adv. Exp. Med. Biol., 112:505.

di Zerega, G.S., and Hodgen, G.D., 1980a, The primate ovarian cycle: suppression of human menopausal gonadotropin-induced follicular growth in the presence of the dominant follicle, J. Clin. Endocrinol. Metab., 50:819.

di Zerega, G.S., and Hodgen, G.D., 1980b, Cessation of folliculogenesis during the primate luteal phase, J. Clin. Endocrinol. Metab., 51:158.

di Zerega, G.S., Nixon, W.E., and Hodgen, G.D., 1980, Intercycle serum FSH elevations: significance in recruitment and selection of the dominant follicle and assessment of corpus luteum normalcy, J. Clin. Endocrinol. Metab., 50:1046.

Evans, W.S., Rogol, A.D., MacCod, R.M., and Thorner, M.O., 1980, Dopaminergic mechanisms and LH secretion. I. Acute administration of the dopamine agonist bromocryptine does not inhibit LH release in hyperprolactinemic women, J. Clin. Endocrinol. Metab., 50:103.

Fuxe, K., Löfsröm, A., Hökfelt, T., Ferland, L., Andersson, K., Agnati, L., Eneroth, P., Gustafsson, J.P., and Skett, P., 1978, Influence of central catecholamines on LHRH-containing pathways, Clin. Obstet. Gynecol., 5:251.

Gautray, J.P., Jolivet, A., Robyn, C., Colin, M.C., and Eberhard, A., 1977, Investigation neuroendocrinienne de la phase folliculaire, au cours de cycles spontanés et après stimulation oestrogènique, Ann. Endocrinol.(Paris), 38:371.

Genazzini, A.R., Camanni, F., Massara, F., Picciolini, E., Cocchi, D., Belforte, L., and Muller, E.E., 1980, A new pharmacological approach to the diagnosis of hyperprolactinemic states: the nomifensine test, Acta Endocrinol., 93:139.

Hanker, J.P., and Schneider, H.P.G., 1980, Pulsatile LH-RH substitution in hypothalamic amenorrhea, Acta Endocrinol., [Suppl.]234:74.

Hillier, S.G., van den Boogaard, A.M.J., Reichert, L.E., and van Hall, E.V., 1980, Intraovarian sex steroid hormone interactions and the regulation of follicular maturation: aromatization of androgens by human granulosa cells in vitro, J. Clin. Endocrinol. Metab., 50:640.

Hoff, J.D., Lasley, B.L., Wang, E.F., and Yen, S.S.C., 1977, The two pools of pituitary gonadotropin: regulation of the menstrual cycle, J. Clin. Endocrinol. Metab., 44:302.

Jaffe, R.B., and Keye, W.R., 1974, Estradiol augmentation of pitui-
     tary responsiveness to gonadotropin releasing hormone in
     women, J. Clin. Endocrinol. Metab., 39:850.
Judd, S.J., Rakoff, J.S., and Yen, S.S.C., 1978, Inhibition of
     gonadotropin and prolactin release by dopamine: effect of
     endogenous estradiol levels, J. Clin. Endocrinol. Metab.,
     47:494.
Knobil, E., 1974, On the control of gonadotropin secretion in the
     rhesus monkey, Recent Progr. Horm. Res., 30:1.
Knobil, E., 1980a, The neural control of the menstrual cycle,
     Acta Endocrinol., [Suppl.]234:154.
Knobil, E., 1980b, The role of signal pattern in the hypothalamic
     control of gonadotropin secretion, in: " Rythmes et Repro-
     duction," Masson et Cie, Paris, p. 73.
Knobil, E., Plant, T.M., Wildt, L., Belchetz, P.E., and Marshall,
     G., 1980, Control of the rhesus monkey menstrual cycle: per-
     missive role of hypothalamic gonadotropin-releasing hormone,
     Science, 207:1371.
Krey, L.C., Butler, W.R., and Knobil, E., 1975, Surgical disconnec-
     tion of the medial basal hypothalamus and pituitary function
     in the rhesus monkey. I. Gonadotropin secretion, Endocrinology,
     96:1073.
Lachelin, G.C.L., Leblanc, H., and Yen, S.S.C., 1977, The inhibitory
     effect of dopamine agonists on LH release in women, J. Clin.
     Endocrinol. Metab., 44:728.
Leyendecker, G., Struve, T., Nocke, W., and Hansmann, M., 1980,
     The permissive role of Gn-RH in the regulation of the human
     menstrual cycle. Pathophysiologic aspects and therapeutic
     consequences, Acta Endocrinol., [Suppl.]234:157.
Leblanc, H., Lachelin, G.C.L., Abu-Fadil, S., and Yen, S.S.C., 1976,
     Effect of dopamine infusion on pituitary hormone secretion in
     humans, J. Clin. Endocrinol. Metab., 43:688.
Lindner, H.R., 1979, Mechanism and significance of luteal desensi-
     tization, Adv. Exp. Med. Biol., 112:703.
Louvet, J.P., Harman, S.M., Schreiber, J.R., and Ross, G.T., 1975,
     Evidence for a role of androgens in follicular maturation,
     Endocrinology, 97:366.
McNatty, K.P., 1979a, Follicular determinants of corpus luteum
     function in the human ovary, Adv. Exp. Med. Biol., 112:465.
McNatty, K.P., 1979b, Relationship between plasma prolactin and
     the endocrine microenvironment of the developing human antral
     follicle, Fertil. Steril., 32:433.
McNeilly, A.S., and Chard, T., 1974, Circulating levels of prolactin
     during the menstrual cycle, Clin. Endocrinol., 3:105.
Meites, J., Bruni, J.F., Van Vugt, D.A., and Smith, A.F., 1979,
     Relation of endogenous opioid peptides and morphine to neuro-
     endocrine functions, Life Sci., 24:1337.
Modlinger, R.S., Schonmuller, J.M., and Arora, S.P., 1980, Adeno-
     corticotropin release by tryptophan in man, J. Clin. Endo-
     crinol. Metab., 50:360.

Morley, J.E., Baranetsky, N.G., Wingert, T.D., Carlson, H.E.,
    Hershman, J.M., Melmed, S., Levin, S.R., Jamison, K.R.,
    Weitzman, R., Chang, R.J., and Varner, A.A., 1980, Endocrine
    effects of naloxone-induced opiate receptor blockade, J. Clin.
    Endocrinol. Metab., 50:251.
Muhlenstedt, D., Bonnet, H.G., Hanker, J.P., and Schneider, H.P.G.,
    1978, Short luteal phase and prolactin, Int. J. Fertil.,
    23:213.
Nathan, R.S., Tabrizi, M.A., Halpern, F.S., and Sachar, E.J.,
    1980, Effect of cyproheptadine and atropine on the diurnal
    PRL responses to insulin induced hypoglycemia in normal man,
    J. Clin. Endocrinol. Metab., 51:90.
Niswender, G.D., Menon, K.M.J., and Jaffe, R.B., 1972, Regulation
    of the corpus luteum during the menstrual cycle and early
    pregnancy, Fertil. Steril., 23:432.
Ondo, J.G., 1974, Gamma-aminobutyric acid effects on pituitary
    gonadotropin secretion, Science, 186:738.
Peillon, F., Vincens, M., Cesselin, F., Doumita, R., and Mowszowicz,
    I., 1979, Cycles anovulatoires avec normoprolactinemie
    apparente. Traitment par bromocryptine, Nouv. Press Med.,
    8/40:3269.
Powell, W.S., Hammarström, S., and Samuelson, B., 1974, Prostaglan-
    din $F_{2\alpha}$ receptor in human corpus luteum, Lancet, 1:1120.
Quigley, M.E., and Yen, S.S.C., 1980, The role of endogenous
    opiates on LH secretion during the menstrual cycle, J. Clin.
    Endocrinol. Metab., 51:1979.
Ross, G.T., and Hillier, S.G., 1978, Luteal maturation and luteal
    phase defect, Clin. Obstet. Gynecol., 5:391.
Selmanoff, M.K., Pramik-Holdaway, M.J., and Weiner, R.I., 1976,
    Concentration of dopamine and norepinephrine in discrete
    hypothalamic nuclei during the rat estrous cycle, Endocrin-
    ology, 99:326.
Seppala, M., Hirvonen, E., and Rauta, T., 1976, Hyperprolactinemia
    and luteal insufficiency, Lancet, 1:229.
Shaar, C.J., Frederickson, R.C.A., Dininger, N.B., and Jackson, L.,
    1977, Enkephalin analogues and naloxone modulate the release
    of GH and PRL. Evidence for regulation by an endogenous
    opioid peptide in brain, Life Sci., 21:853.
Shemesh, M., and Hansel, W., 1975, Stimulation of prostaglandin
    synthesis in bovine ovarian tissues by arachidonic acid and
    LH, Biol. Reprod., 13:448.
Singh, E.J., Baccarini, I.M., and Zuspan, F.P., 1975, Levels of
    prostaglandins $F_{2\alpha}$ and E_2 in human endometrium during the
    menstrual cycle, Am. J. Obstet. Gynecol., 121:1003.
Takahara, J., Yunoki, S., Yakushiji, W., Yamauchi, J., Yamane, Y.,
    and Ofuji, T., 1977, Stimulatory effects of Gamma-hydroxy-
    butyric acid on growth hormone and prolactin release in
    humans, J. Clin. Endocrinol. Metab., 44:1014.

Thibault, C., and Levasseur, M.C., 1979, La Fonction Ovarienne Chez les Mammifères," Masson et Cie, Paris.

Tyson, J.E., and Pinto, H., 1978, Identification of the possible significance of prolactin in human reproduction, Clin. Obstet. Gynecol., 5:411.

Vande Wiele, R.L., and Turksoy, R.N., 1965, Treatment of amenorrhea and of anovulation with human menopausal and chorionic gonadotropins, J. Clin. Endocrinol. Metab., 25:369.

Van Loon, G.R., Ho D., and Kim, C., 1980, Beta-endorphin-induced decrease in hypothalamic dopamine turnover, Endocrinology, 106:76,

Vekemans, M., Delvoye, P., L'Hermite, M., and Robyn, C., 1977, Serum prolactin levels during the menstrual cycle, J. Clin. Endocrinol. Metab., 44:1222.

Wang, C.F., Lasley, B.L., Lein, A., and Yen, S.S.C., 1976, The functional changes of pituitary gonadotrophs during the menstrual cycle, J. Clin. Endocrinol. Metab., 42:718.

Westwerdt, W., Muller, O., Brandau, H., 1979, Structural analysis of granulosa cells from human ovaries in correlation with function, Adv. Exp. Med. Biol.., 112:123.

Wildt, L., Marshall, G., and Knobil, E., 1980, Experimental induction of puberty in the infantile female rhesus monkey, Science, 207:1373.

Yen, S.S.C., 1978, The human menstrual cycle (Integrative function of the hypothalamic-pituitary-ovarian-endometrial axis), in: "Reproductive Endocrinology," S.S.C. Yen and R.B. Jaffe, ed., W.B. Saunders Co., Philadelphia, p. 126.

Yen, S.S.C., and Lein, A., 1976, The apparent paradox of the negative and positive feedback control system on gonadotropin secretion, Am. J. Obstet. Gynecol. 126:942.

Yen, S.S.C., and Tsai, C.C., 1972, Acute gonadotropin release induced by exogenous estradiol during the midfollicular phase of the menstrual cycle, J. Clin. Endocrinol. Metab., 34:298.

Yen, S.S.C., VandenBerg, G., Rebar, R., and Ehara, Y., 1972, Variation of pituitary responsiveness to synthetic LRF during different phases of the menstrual cycle, J. Clin. Endocrinol. Metab., 35:931.

Young, J.R., and Jaffe, R.B., 1976, Strength duration characteristics of estrogen effects on gonadotropin response to gonadotropin releasing hormone in women. II. Effects of varying concentrations of estradiol, J. Clin. Endocrinol. Metab., 42:432.

SURFACE CHANGES OF THE LUMINAL UTERINE EPITHELIUM DURING THE

HUMAN MENSTRUAL CYCLE: A SCANNING ELECTRON MICROSCOPIC STUDY

Dominique Martel, Catherine Malet, Jean Pierre Gautray,
and Alexandre Psychoyos

Laboratoire de Physiologie de la Reproduction
CNRS, E.R. 203
Hopital de Bicetre
78 Av du General Leclerc
94270 Bicetre, France

INTRODUCTION

    The two ovarian hormones, estradiol and progesterone, induce
profound morphologic changes in the endometrium. Classic histological techniques enabled detailed description of these alterations very early and extensive studies have permitted the dating
of endometrial biopsy and precise pathology (Noyes et al., 1950;
Gautray, 1968). According to these studies, the most relevant
endometrial changes deal with stromal reorganization; however,
more recently, electron microscopic studies have revealed subtle
changes in the luminal surface epithelium (for reviews see Borell
et al., 1959; Gompel, 1962; Johannisson and Nilsson, 1972; Nilsson
and Nygren, 1972; Ferenczy and Richart, 1973; Wynn, 1977).

    Study of the changes in the uterine luminal surface is of
interest since epithelial cells are directly implicated in
biological processes related to uterine secretion, spermatozoa
transport, and egg implantation. Furthermore, recent findings in
our laborotory (Martel and Psychoyos, 1980) show that epithelial
and stromal cells of the rat uterus are affected differently by
the same hormonal sequence. As shown by parallel transmission
and scanning electron microscopic studies (Nilsson and Nygren,
1974; Wynn, 1977), the features of the epithelial cell surface
reflect some important functional changes. This paper deals
with supplementary information concerning the ultrastructural
surface changes of the luminal epithelium of the human uterus and

15

attempts to find criteria for dating endometrial biopsies by
scanning electron microscopy.

METHODS

The data reported in this paper are based on human endometrial biopsies; only samples that appeared normal upon histological examination were included in the study. Specimens were
washed in physiologic serum immediately after surgery, fixed in
2.5% glutaraldehyde, postfixed in 1% osmium textroxide, dehydrated
with acetone and dried in a critical-point drier with carbon
dioxide. They were then mounted, coated with gold palladium and
examined in a JEOL SM 75 type microscope.

RESULTS AND DISCUSSION

The luminal epithelium is composed of two cell types: micro-
villi-bearing cells and ciliated cells. The surface of the ciliated cells is covered with kinocilia, which presumably promote the
flow of uterine fluids, thus favoring repartition of uterine
secretions, movement of spermatozoa and migration of the ova. Our
results did not reveal any major change in ciliated cell frequency
throughout the menstrual cycle even though the ciliated cells were
expected to be more numerous in the first part of the cycle because
of stimulation by estradiol (Flemming et al., 1968; Ferenczy et al.,
1972; Masterson et al., 1975). Because the ciliated cells are not
uniformly spread along the uterus (Schueller, 1968; Ferenczy et al.,
1972; Johannisson and Nilsson, 1972), we consider the ratio of
ciliated cells to nonciliated cells a random criterion for biopsy
dating. The prominent cell type of the luminal epithelium is
the microvillous cell. Its apical surface appears to be evolving
constantly, depending on a specific hormonal effect. A single
observation of the appearance of these cells may thus indicate
the general endometrial hormonal status.

During the early proliferative phase, the endometrial epithelium regenerates from glandular cells remaining after the preceding
menstrual cycle. Under the influence of estrogens, the two epithelial cell types proliferate. Figure 1 shows ciliated cells at
various stages of growth; the microvillous cells appear fusiform
at this stage, reflecting intense proliferation. The microvilli
are numerous, thin and long (Fig. 1A, B). By the mid proliferative
phase (Fig. 2) reepithelization is achieved and ciliated cells
reach their maximum size. At this stage, the microvillous cells
have a polyhedral shape with longer, dense, flat microvilli.

Near the ovulatory period (Fig. 3) the first striking change
in the microvillous cells is visible. The cell surface appears to
bulge, the microvilli diminish in size, become thick, and project
upward from the surface of the cell toward the uterine lumen. Just
after ovulation (days 14 to 16 of the cycle), numerous droplets

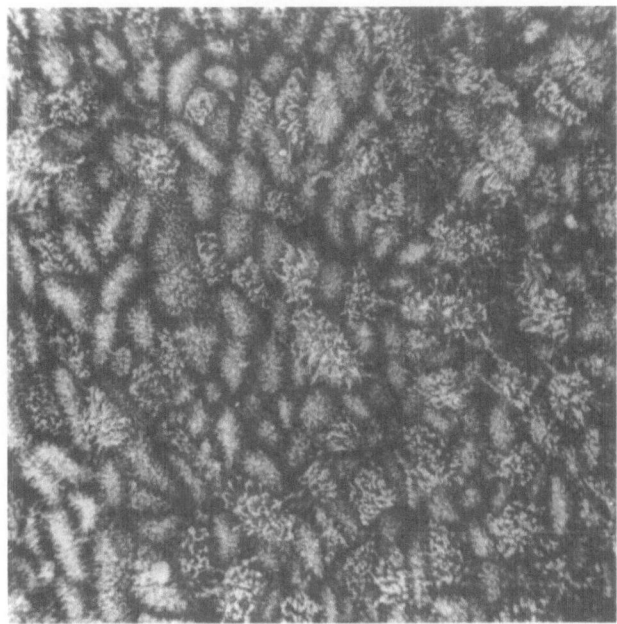

Fig. 1A. Luminal epithelium of the human uterus during the early
          proliferative phase (x 700).

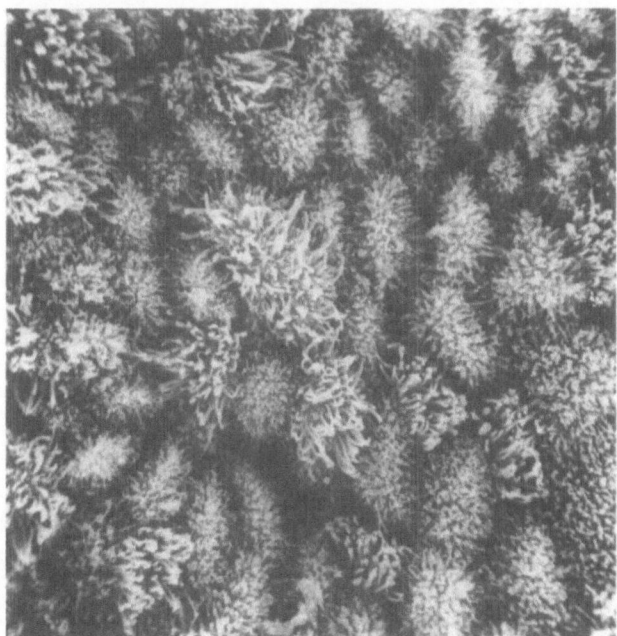

Fig. 1B. Luminal epithelium of the human uterus during the early
          proliferative phase (x 1450).

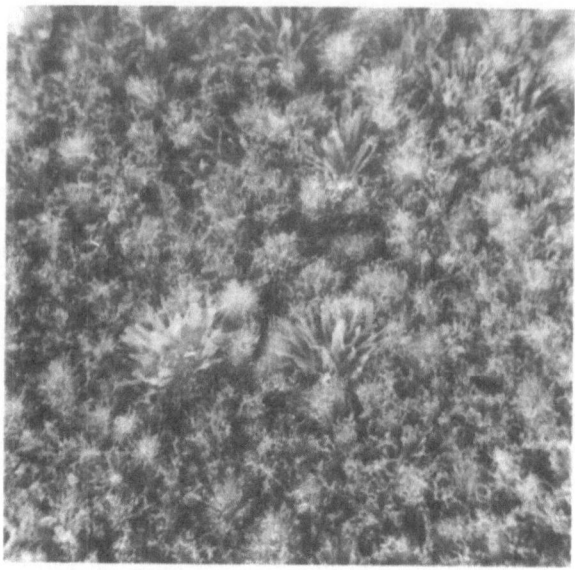

Fig. 2. Luminal epithelium during the mid proliferative phase
         (x 1450).

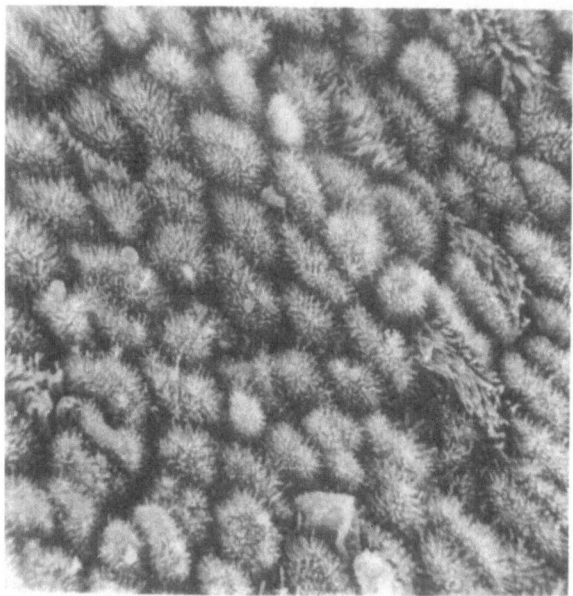

Fig. 3. Luminal epithelium during the ovulatory period (x 1450).

cover the luminal surface (Fig. 4A,B) and may correspond to a
secretory product. From histological studies, the vacuolization
of the epithelium is maximal at day 16. The fact that nuclei are
first pushed upward in the cell and then, after the vacuoles have
passed, return to their normal position, indicates a secretory
process.

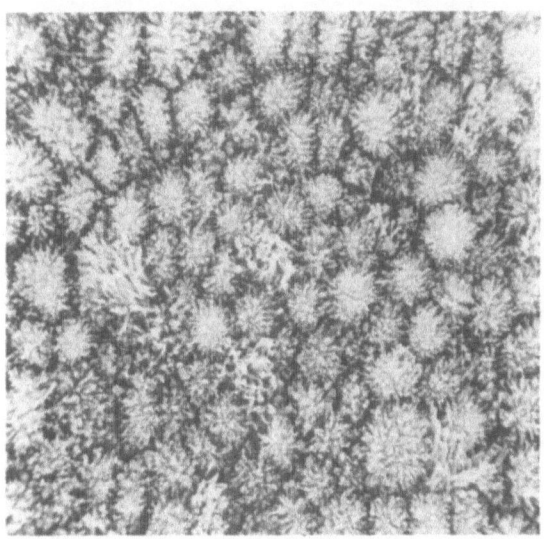

Fig. 4A. Luminal epithelium of the human uterus just after ovu-
          lation (days 14 to 16 of the cycle) x 1450.

Fig. 4B. Luminal epithelium of the human uterus just after ovu-
          lation (days 14 to 16 of the cycle) x 2900.

In samples removed at days 16 to 18 of the cycle (Fig. 5A,B),
the droplets that appeared just after ovulation are still present
and the swelling of the microvillous cells is more prominent. By
days 18 to 19 of the cycle, ectoplasmic projections are clearly
distinguishable on the cell surface and give the microvillous

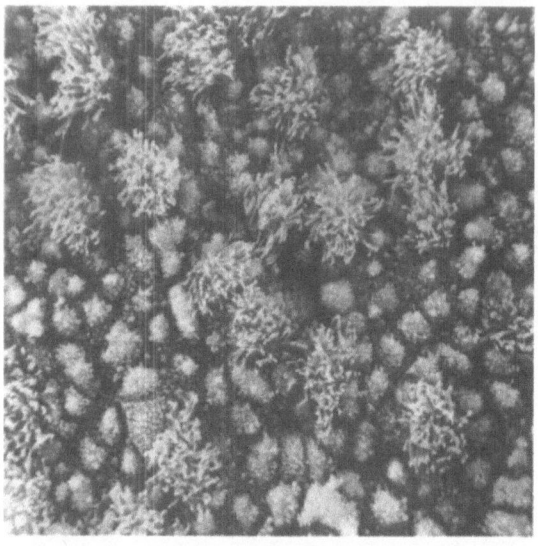

Fig. 5A.   Luminal epithelium on days 17 to 19 of the cycle (x 1450).

Fig. 5B.   Luminal epithelium on days 17 to 19 of the cycle (x 3500).

cells the appearance of sea anemones (Fig. 6A). As the formation
of these structures continues, the microvilli become progressively
smaller and thicker, as illustrated in Fig. 6B. The microvilli
then completely disappear and the sea anemone-like projections
appear completely developed on days 20 to 21 of the cycle. At this

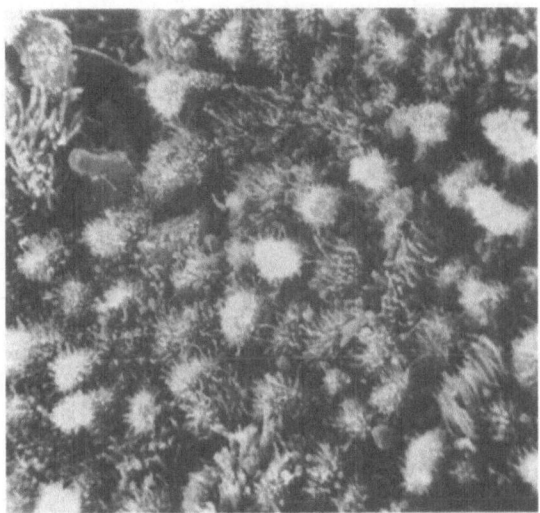

Fig. 6A. Luminal surface epithelium at the time of ovoimplantation.
Microvillous cells appear swollen and resemble sea
anemones (x 1450).

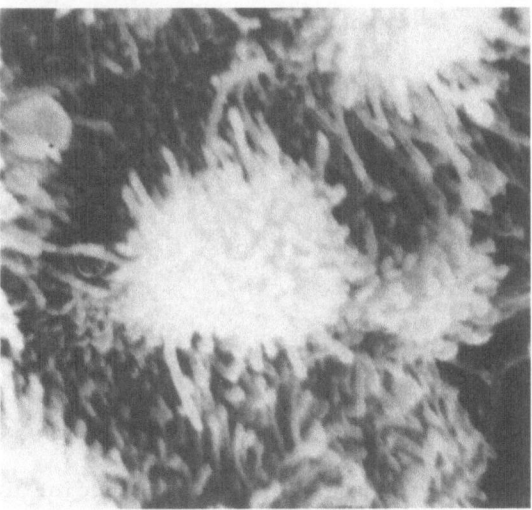

Fig. 6B. Luminal surface epithelium at the time of ovoimplantation
Microvillous cells appear swollen and resemble sea
anemones (x 7000).

time, as shown in Fig. 7, the entire surface of the cells appear
to be covered by a folded structure resembling a sponge. The
period during which these structures are visible is limited to 24
to 48 hours. By days 22 to 23 of the cycle, they have largely
regressed (Fig. 8).

    Similar ectoplasmic projections of the apical surface of the
uterine epithelial cells have been observed by scanning electron
microscopy and were first described in the rat by Psychoyos and
Mandon (1971a,b). In the rat, these projections present during
delayed implantation become abundant on the day of initiation of
egg implantation, and disappear thereafter (Psychoyos, 1973). It
can, therefore, be suggested that in the human, as in the rat, the
presence of these projections is restricted to the perinidatory
period. Psychoyos and Mandon (1971a,b) in discussing the eventual
function of these projections, mentioned the early studies of Vokaer
and Leroy (1962), in which trypan blue was transported from the
uterine lumen into the epithelial cells and stroma, and suggested
the possibility that these projections are involved in the uptake
of substances (athrocytosis) and luminal fluid (pinocytosis). In

Fig. 7.   By days 20 to 21, the ectoplasmic projection process
          is maximal, microvilli disappear and the cell surface
          has a sponge-like appearance (x 1450).

Fig. 8. By days 22 to 23, structures regress (x 3500).

fact, it is now evident that these projections, termed pinopodes, are actively involved in the endocytosis of uterine fluids (Enders and Nelson, 1973) and various macromolecules (Parr and Parr, 1974; Parr, 1980). Their presence at the time of egg implantation may certainly aid in the withdrawal of uterine fluid and/or in several phenomena, i.e., apposition of the blastocyst to the luminal epithelium or the interaction between the blastocyst and the stroma.

The uterine epithelium regains a regular appearance on day 23 of the cycle (Fig. 9). The nonciliated cells exhibit a polygonal shape, and the microvilli resume the characteristic appearance of the second part of the cycle, i.e., they are less numerous, small and thick. By day 25 of the cycle (Fig. 10), the microvillous cells become dome shaped, and are distinctly separated from each other, with scattered tiny microvilli that appear to be small points on the cell surface. These pseudodecidual cells (Fig. 10) resemble those of true decidual tissue (Fig. 11A,B); however, in this latter case, the cells are elongated, which reflects intense proliferation. In addition, in the true decidual reaction, the ciliated cells completely disappear, contrary to what is observed in pseudodecidual tissue. There is no major change from day 25 to day 28 of the cycle.

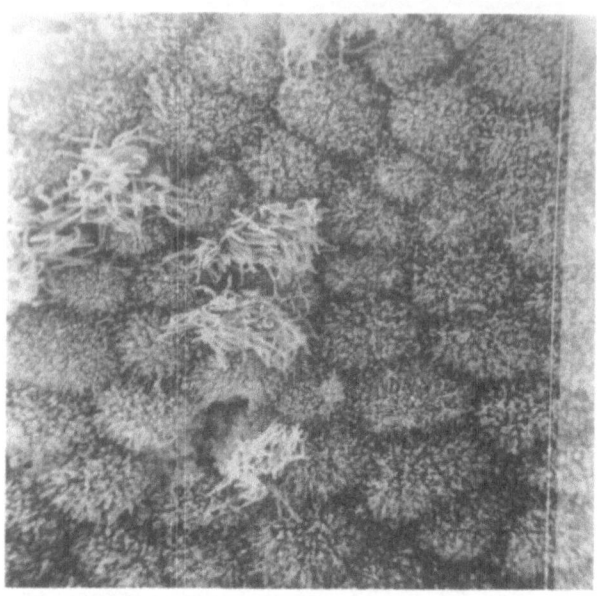

Fig. 9.   Luminal surface epithelium by days 23 to 24 of the cycle
          (x 1450).

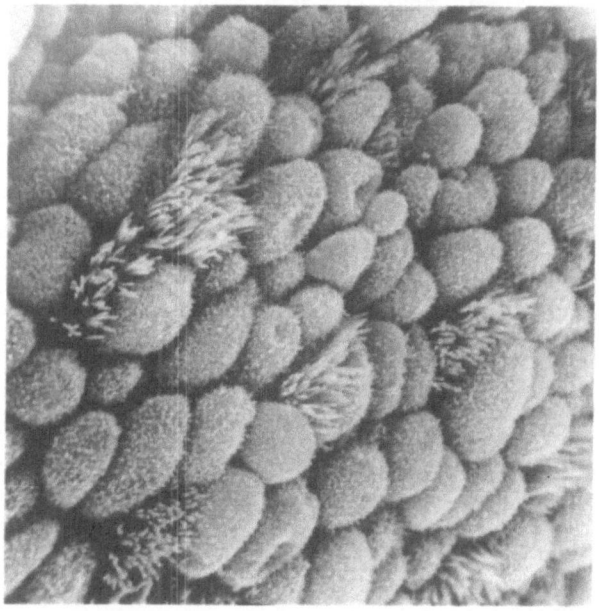

Fig. 10.   Pseudodecidual reaction (x 1450).

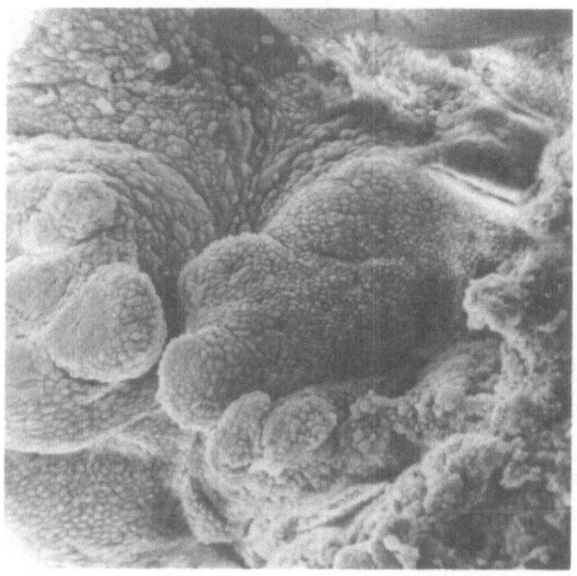

Fig. 11A. Appearance of the epithelium of decidual tissue (x 150).

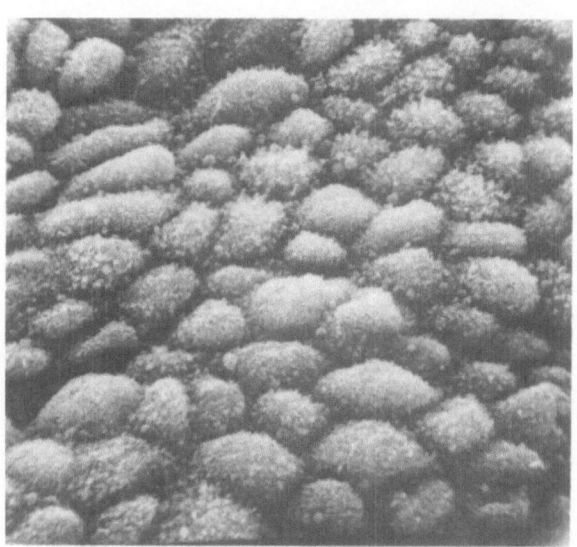

Fig. 11B. Appearance of the epithelium of decidual tissue
(x 1450).

Figure 12 shows an hypotrophic endometrium, as determined
by histological examination. The biopsy was performed on day 23
of the menstrual cycle. Observation of this sample under the
scanning electron microscope revealed that the microvillous cells
with their numerous long thin microvilli correspond to those of a
mid-proliferative phase. Furthermore, it can be seen that regeneration of the epithelium is not achieved since stromal cells are
visible. Interestingly, apical projections are present that
correspond to days 20 to 21 of the cycle and are compatible with
the date of biopsy.

A hyperplastic endometrium removed on day 20 of the cycle
(Fig. 13), shows tiny scattered microvilli that correspond to
days 24 to 25 of the cycle. However, under hyperestrogenic stimulation, the microvillous cells undergo an altered growth process,
as reflected by their irregular, elongated shape and considerably
enlarged size. The observation of ciliary buds and of ciliated
cells at different stages of development indicates active cilia
formation, which is incompatible with the second part of the cycle.

CONCLUSION

Scanning electron microscopic observation of the luminal
epithelium shows the continuous evolution of the ultrastructure
of the cell surface that escapes classic investigation technique.

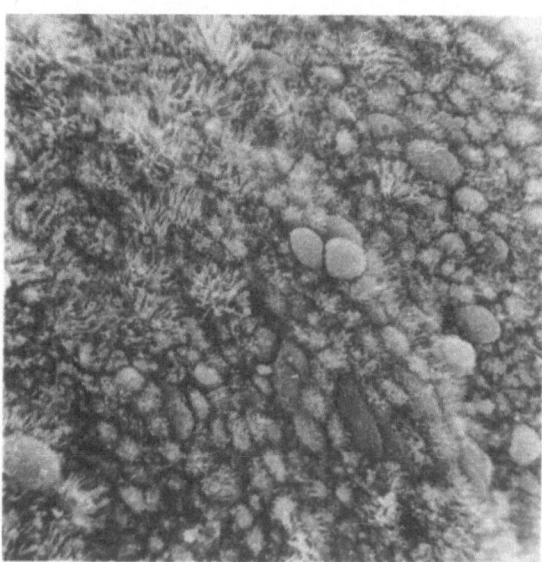

Fig. 12. Luminal surface epithelium of hypotrophic endometrium
          (x 1450).

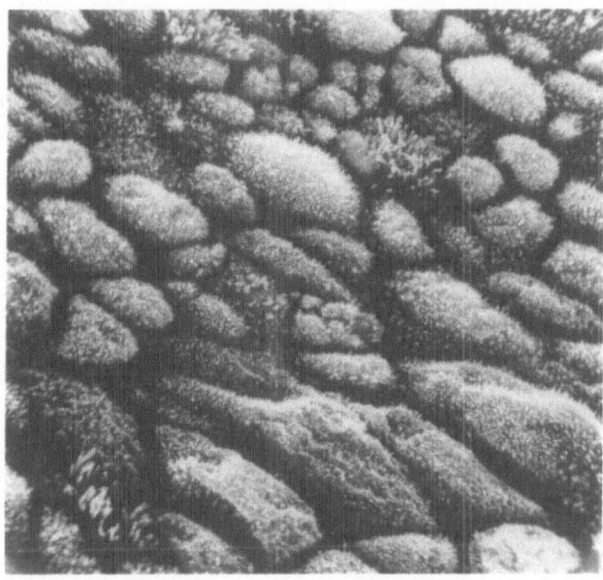

Fig. 13. Hyperplastic endometrium (x 1450).

The two cellular types that compose the luminal epithelium, ciliated and microvillous cells, proliferate under the influence of
estrogen. There are no major alterations in the appearance of the
ciliated cells throughout the cycle, whereas the surface of the
microvillous cells is constantly evolving and reflects the successive functional states of the cell: proliferation, secretion,
pinocytosis, pseudodecidualization and degeneration. These changes
are well defined in time, especially during the second part of the
cycle. Thus, by observing the uterine epithelial surface, it is
possible to date an endometrial biopsy within 2 to 3 days and to
obtain information concerning an eventual uterine dysfunction
caused by hormonal imbalances. Scanning electron microscopy of
the endometrium may thus provide a simple and useful tool for
clinicians to supplement information obtained by usual clinical
techniques.

Acknowledgments

     The authors are thankful to Mrs. Guillaumin for help and
advice concerning scanning electron microscopy. We wish also
to thank the laboratoire d'évolution des êtres organisés de
l'Université Paris VI for giving us the opportunity to use the
scanning electron microscope.

     This work was supported by the Centre National de la Recherche
Scientifique, and the Délégation Générale à la Recherche Scientifique et Technique.

REFERENCES

Borell, U., Nilsson, O., and Westman, A., 1959, The cyclical
    changes occurring in the epithelium lining the endometrial
    glands. An electron microscopical study in the human being,
    Acta Obstet. Gynecol. Scand., 38:364.
Enders, A.C. and Nelson, D.M., 1973, Pinocytotic activity in the
    uterus of the rat, Am. J. Anat., 138:277.
Ferenczy, A., Richard R.M., Agate, F.J., Jr., Purkerson, M.L.,
    and Dempsey, E.W., 1972, Scanning electron microscopy of
    the human endometrial surface epithelium, Fertil. Steril.,
    23:515.
Ferenczy, A., and Richart, R.M., 1973, Scanning electron micro-
    scopy of human endometrial surface epithelium, J. Clin.
    Endocr. Metab., 36:999.
Flemming, F., Tweeddale, D.N., and Roddick, J.W., 1968, Ciliated
    endometrial cells, Am. J. Obstet. Gynecol., 102:186.
Gautray, J.P., 1968, Reproduction humaine aspects actuels de biol-
    ogie clinique, Masson, Paris.
Gompel, C., 1962, The ultrastructure of the human endometrial
    cells studied by electron microscopy, Am. J. Obstet.
    Gynecol., 84:1000.
Johannisson, E., and Nilsson, L., 1972, Scanning electron micros-
    copy of the human endometrium, Fertil. Steril., 23:613.
Martel, D., and Psychoyos, A., 1980, Behaviour of uterine steroid
    receptors at implantation, Prog. Reprod. Biol., 7:216.
Masterson, R., Armstrong, E.M., and More, I.A.R, 1975, The cyclical
    variation in the percentage of ciliated cells in the normal
    human endometrium, J. Reprod. Fertil., 42:537.
Nilsson, O., and Nygren, K.G., 1972, Scanning electron microscopy
    of the human endometrium, Ups. J. Med. Sci., 77:3.
Nilsson, O., and Nygren, K.G., 1974, Ultrastructure of human
    uterine epithelium at the time of implantation after post-
    ovulatory administration of norethindrone, Ups. J. Med. Sci.,
    79:65.
Noyes, R.W., Hertig, A.T., and Rock, J., 1950, Dating the endome-
    trial biopsy, Fertil. Steril., 1:3.
Parr, M.B., and Parr, E.L., 1974, Uterine luminal epithelium
    protrusions mediate endocytosis, not apocrine secretion
    in the rat, Biol. Reprod., 11:220.
Parr, M.B., 1980, Endocytosis in the uterine epithelium during
    early pregnancy, Prog. Reprod. Biol., 7:81.
Psychoyos, A., 1973, Hormonal control of ovoimplantation, Vitam.
    Horm., 31:201.
Psychoyos, A., and Mandon, P., 1971a, Etude de la surface de
    l'épithélium utérin au microscope électronique à balayage.
    Observation chez la ratte au 4ème et 5ème jours de la
    gestation, C.R. Acad. Sci. Paris, 272:2723.

Psychoyos, A., and Mandon, P., 1971b, Scanning electron micros-
    copy of the surface of the rat uterine epithelium during
    delayed implantation, J. Reprod. Fertil., 26:137.
Schueller, E., 1968, Ciliated epithelioid of the human uterine
    mucosa, Obstet. Gynecol., 31:215.
Vokaer, R., and Leroy, F., 1962, Experimental study of focal
    factors in the process of ovoimplantation in the rat, Am.
    J. Obstet. Gynecol., 83:141.
Wynn, R., 1977, Histology and ultrastructure of the human endo-
    metrium, in: "Biology of the Uterus," R. Wynn, ed., Plenum
    Press, New York.

ANALYSIS OF ISOLATED AND COMBINED ACTIONS

OF OVARIAN STEROIDS ON THE ENDOMETRIUM

Jean de Brux

Institut de Pathologie et de Cytologie Appliquee
53, rue des Belles-Feuilles
75116 Paris (F)

Discovery of intracellular hormonal receptors has permitted histologic interpretation of the action of ovarian steroids on the endometrium. Research carried out by biochemists and physiologists has led to a better understanding of the histologic effects of hormonal disturbances and isolated or combined hormonal treatments.

The constituents of the endometrium retain their embryonic potential, but their responses to steroids are not necessarily identical due to their functional differentiation; hence, the diversity of these responses to various pure or synthetic hormones (singly or in combination) and to the time of their administration. Therefore, we examined the localization and efficacy of the receptors of each constituent, the effects of the isolated hormones on each of them and cellular responses to various combinations of hormones.

LOCALIZATION AND CHARACTERISTICS OF HORMONE RECEPTORS

Steroids in the blood penetrate intercellular spaces (where steroid levels are high) following the capillary beds, which filter the hormones. Two steroid types are distinguished: the free steroids, and those that bind to intercellular proteins at the cytoplasmic membrane, a true estrogen receptor. Clark et al. (1978b) have isolated two cytoplasmic receptors, but acknowledge that others may exist.

Cytosolic Estrogen Receptors

Clark et al. (1978b) found that 50% of plasma steroids are free. This percentage is accepted by the classic intracytosolic receptor, or Type I receptor, which is composed of two "subunits": subunit α, a macromolecule of 80,000 daltons and subunit β, a macromolecule of 50,000 daltons. Both subunits probably move to the nucleus by translocation, are not found in the cytoplasm after injection of estrogens and seem to have a strong affinity for estrogens, but only a small capacity for accumulation.

The remaining 50% of the steroids, associated with the lipids, proteins and serum albumin, concentrate on the periphery of the cell in the intercellular spaces. The cytoplasmic membrane thus constitutes the Type II receptor, which has a weak affinity for hormones, but a great capacity for accumulation. This complex, Receptor ^{E} Type II, penetrates the cytoplasm by endocytosis, but does not translocate to the nucleus. It seems to represent a precursor of the Type I receptor after translocation, and to constitute a reserve of weak-affinity macromolecules.

Thus, the Type II receptor accumulates bound estrogens, which are gradually absorbed according to need, i.e., when the level of free estrogens is reduced or absent. These Type II receptor-complexes thus act as a "steroid reservoir", whose absorption by the cell may be compared with that of cholesterol-lipoprotein complexes.

Nuclear Receptors

The receptor-estrogen complex, after translocation across the nuclear membrane, binds to two types of nuclear acceptors: the specific chromatin acceptor, which has a strong affinity for the Receptor^E complex; and the nonspecific acceptor, situated on interchromatin substances. Their difference has been demonstrated by the ease with which the nonspecific acceptor may be extracted by solubilization in highly concentrated saline solution.

Renewal or recycling of cytosolic and extracytosolic receptors is stimulated by linking receptor^E with acceptor^N. Thus, nuclear retention of the receptor^E complex may determine an irreversible mechanism that stimulates transcription, followed by biosynthesis leading to cellular hypertrophy and hyperplasias of the target-cells. The process can be stopped only by hormone suppression or blocking the receptors.

Progesterone Receptors

Progesterone receptors are induced by estrogens. In guinea pigs, the concentration of progesterone receptors is determined by the degree of estrogen accumulation in endometrial cells

(Warembourg and Milgrom, 1977). Estrogens, thus control the synthesis of progesterone receptors. The linking of cytosolic progesterone receptors with nuclear acceptors is followed by the disappearance of cytoplasmic progesterone receptors, which reappear
when the plasma estrogen level again becomes sufficient to resume
synthesis of progesterone receptors. Thus, progesterone controls
its own receptors.

Progesterone also controls the level of estrogen receptors,
and, therefore has an antagonistic action on estrogens since it
hinders reformation of cytosolic estrogen receptors (Means and
O'Malley, 1971; Milgrom et al., 1973; Brenner et al., 1974; Pavlik
and Couson, 1976). By this mechanism, progesterone reduces or
inhibits estrogen stimulation of the uterus (Hsueh et al., 1976).

Action of Estrogens on Genital Tract

Numerous investigators have studied the mechanism by which the
steroid hormones stimulate cellular functions (Jensen et al., 1973,
1974; Baulieu et al., 1975; O'Malley and Means, 1974; Gorski and
Gannon, 1976; Yamamoto and Alberts, 1976). Clark and co-workers
(1978a) injected estrogens dissolved in saline solution into rats
to study the direct and isolated action of the hormone.

Response to Combination Cytoplasmic Estrogen Receptor (RCE)-Nuclear Receptor (RCN)

The quantity of RCE-complex that accumulates in the nucleus
is proportional to the quantity of estradiol injected, and the
concentration is maximal at the end of 1 hour. However, nuclear
retention of the RCE-complex is not proportional to the dose of
estrogens injected. In fact, even if all nuclear sites (15,000
to 20,000) are saturated, 80% disappear within 8 hours. However,
if estrogens are injected in small quantity, the RCE-complexes
remain within the nucleus for more than 6 hours and the cell
establishes its own metabolic balance.

Uterine responses to estrogens may be separated into early
and late responses. Early responses are characterized by an increase in RNA polymerase, aqueous imbibition, histamine mobiliza-
tion, hyperemia, formation of cyclic AMP, transitory increases in
RNA, lipid, and protein synthesis, and increase in and synthesis of
sugars. Late responses are characterized by continuous RNA synthesis, protein synthesis (e.g., specific uteroglobulin), and
cellular growth (hyperplasia, mitoses). These responses are due
to stimulation of the chromatin sites.

VARIABILITY OF ESTROGEN ACTION

Estradiol

Estrogens vary in structure and activity. Hisaw (1959), Huggins and Jensen (1955) and Wotiz et al. (1968) consider estradiol a weak estrogen, because it remains in the receptor only a short time; however, it blocks certain nuclear receptors, thus preventing the reconstitution of cytoplasmic receptors. In ovariectomized animals, separate injection of estriol and estradiol or injection of a mixture of the two, produced increased uterine weight and showed that estriol did not act as an antagonist of estradiol. Therefore, the weak estrogenicity of estriol seems to be linked to its brief occupation of nuclear receptors. However, continuous injection of estriol induces long-term nuclear retention, blocking receptors and preventing entry of other estrogens. Under these circumstances, estriol acts as a true antiestrogen.

Clark and Peck (1979) have classified estrogen antagonists and agonists into two large categories: those with brief activity and those with long activity (Table 1).

Derivatives of Triphenylethylene (TPE)

Nafoxidine, given alone, doubles the volume of the uterus; however, simultaneous estradiol-nafoxidine treatment produces no growth. Clark et al. (1978b) postulate that the cellular stimulation induced by nafoxidine is different for the various target cells, inducing hypertrophy of the surface epithelium and only weak stimulation of the stroma and myometrium. This difference is observed in the entire reproductive system of animals.

Tamoxifen has the same properties. Experimentally, TPE derivatives mixed with diethylstilbestrol (DES) inhibit the action of DES (no uterine growth, absence of glandular tubes), but produce persistent cellular hypertrophy. This substance acts as an estrogen-like stimulant if given alone and an estrogen antagonist if given simultaneously with an estrogen.

These effects occur in all tissues containing steroid receptors, explaining their action in breast cancer. The difference in effect may be explained by the fact that TPE derivatives are retained for long periods by nuclear estrogen receptors and prevent the renewal of cytoplasmic receptors. However, this explanation is perhaps somewhat oversimplified.

When the isolated actions of TPE derivatives and estradiol were studied by electron microscopy, there was a notable increase in cellular height and golgi bodies, and numerous polysomes and

Table 1. Agonists and Antagonists of Estrogens

Varieties	Type	Nuclear Retention	Renewal of Rc	Pharmacological Characteristics	Uterotropic Properties
Short activity	Estriol Dimethylstilbestrol	Brief (1/4 hr)	Positive (within 24 hr)	Partially agonist-antagonist Agonist in continuous treatment	Early responses
Long activity	Estradiol Diethylstilbestrol Ethylestrol Cyclopentyl Ether	Intermediate (6-24 hr)	Positive (within 24 hr)	Agonist	Early and late responses
	Derivatives of Triphenylethylene (nafoxidine) (tamoxifen) (clomid)	Long (more than 24 hr)	Negative (after 24-28 hr)	Antagonist Different cellular stimulation	

From Clark and Peck, 1979.

abundant endoplasmic reticulum were observed. These facts are
evidence of marked cytoplasmic protein synthesis. Nuclear bodies
are found (20%) in the nucleus 72 hours after injection; more are
seen if estrogen stimulation is continuous. Four varieties of
nuclear bodies were noted: filamentous (on the periphery); granular, with clear centers; double; and granular, with central inclusions. Bouteille et al. (1967) and Dupuy-Coin et al. (1972)
demonstrated that these nuclear bodies were composed of a filamentous protein capsule with RNA or DNA at the center and Krishan
et al. (1967) found the same nuclear bodies in certain tumor cells.
They indicate marked transcriptional activity and may be a precursor of carcinoma if estrogen stimulation is very prolonged, and
in fact, are found in endometrial carcinoma cells.

ESTROGENS AND CANCER

In many cases, growth and maintenance of cancer cells depend
on the persistence of estrogen stimuli; some cancers have been
linked to continuous use of estrogens. However, their role in
carcinogenesis has not yet been clearly established (Hertz, 1974).

Lacassagne (1932) demonstrated that rats and mice subjected
to repeated injections of estrogens develop genital and mammary
cancers. It is also known that women under permanent estrogen
treatment present a relatively high proportion of endometrial
cancers or precancerous lesions; however, the estrogen doses are
always moderate, prolonged, uninterrupted and not counterbalanced
by an antagonist. Cellular metabolism, is, thus, constantly aided,
as manifested by: increased protein and RNA synthesis; permanent
stimulation of RNA polymerase action; increase and modification of
proteins; and DNA synthesis and permanent mitoses. These mani-
festations lead to cellular hypertrophy, hyperplasia, growth, and
metabolic disturbances, but do not, however, affirm that estrogens
are the only cause of carcinomas.

ACTION OF PROGESTERONE

During the follicular phase of the cycle, there are relatively
few progesterone receptors (R^P), but as the plasma estrogen level
rises, they increase proportionately (Milgrom et al., 1972; Leavitt
et al., 1978). Thus, R^P is at its maximum in the preovulatory phase
because of the action of estrogen on R^P synthesis. This increase
in R^P is probably necessary for the action of progesterone during
the luteal phase of the cycle and during pregnancy. Later, the
R^P level decreases when the corpus luteum produces progesterone
in great quantity (Pollow et al., 1977).

Estrogens thus increase uterine growth and tissue progesterone receptors, whereas progesterone is an antagonist of estrogens and reduces the number of estrogen receptors. The interaction of the ovarian hormones at the receptor level is the basis for cyclic tissue changes during the menstrual cycle.

MORPHOLOGIC CHANGES OF THE ENDOMETRIUM

After ovariectomy, the weight of the uterus diminishes considerably. The functional zones of the endometrium regress, edema disappears, and stromal cells are transformed into fibroblasts, which deposit fibers of reticulin and then collagen. Basal vessels become stenosed due to thickening and sclerosis of the walls. Glandular tubes cease to multiple and remain quiescent, flattened, or dilated. The cytoplasm loses most of its granulation and sometimes becomes eosinophilic without secretion; the nuclei are small and retracted, chromatin fades, and the nucleoli disappear. The microvilli are sparse and short.

Early Responses to Estrogens

Edema appears within 6 to 10 hours after estrogen injection. Fluid differs from that of the peritoneum in that it is more basic and contains proteins (uteroglobin) and enzymes (ribonuclease, phosphatase, amylase, trypsin). Mucopolysaccharides and acid sulfo-mucopolysaccharides are abundant especially in the stroma, and probably correspond to depolymerization of the ground substance and water retention. This picture resembles that found in the sexual skin of primates.

Edema, provoked by hyperemia, results from an epithelial cell secretion (estrogen induces vacuole formation in the cells) and from capillary growth induced by estrogens (the endo- and perithelial cells are rich in estrogen receptors). The capillaries dilate, pores and intercellular junctions become visible, and walls increase in permeability, thus permitting passage of protein-rich liquid.

Late Response to Estrogens

The epithelial response is manifested by increased cell length due, at first, to edema by hydration of intracellular proteins. Later, the nuclei swell and become irregular with large nucleoli, indicating synthesis activity. The cytoplasm becomes basophilic, the golgi become large and the number of ribosomes and estrogen receptors increases. The microvilli increase in number and size, and are covered with small projections with a glycocalyx on the surface. The glycocalyx, formed in the golgi apparatus and filtered through the apical cytoplasmic membrane, probably acts as support for the alkaline phosphatase (Lawn, 1973).

The stromal response is less well known. The stroma first
appears dissociated because of edema. The cells are generally
elongated or slightly stellate; some show a few mitoses. The
electron microscope shows increased growth and complexity of the
endoplasmic reticulum and golgi apparatus, indicating protein and
connective tissue formation with subsequent deposit of collagen
fibers.

Responses to Progesterone

The action of progesterone alone on the endometria of ovariectomized animals and menopaused women is very weak, as might be
expected from receptor studies. However, after administration of
estrogens followed by progesterone, the epithelial cells widen, round
out, become eosinophilic and uterine "lace" appears at the surface
of the tubes. Their secretion is composed of glycogen, mucopolysaccharides and proteins that are absent in the plasma. The
function of these proteins is probably nourishment of the blastocyst. The stromal response essentially affects the cells; the
nuclei round out, the chromatin disperses, the nucleolus becomes
distinct, and the cytoplasm widens, becomes eosinophilic, and
contains glycogen. These predecidual changes are particularly
apparent at the periphery of the dilated vessels. Electron microscopy shows numerous cytoplasmic inclusions. The presence of lyso-
somes precedes cellular autolysis. Sometimes interstitial edema
appears, and collagen fibers tend to disappear, leaving only reticulin fibers.

COMBINED ACTIONS OF ESTROGENS AND PROGESTERONE

The surface epithelium of the endometrium is very sensitive
to estrogens, i.e., mitoses begin 24 to 30 hours after injection.
This accelerated cell growth is due to the shortening of the G-1
rest phase. Daily estrogen injections or a single injection of
long-acting estrogen stimulates mitosis after 2 days, after which
the response lessens.

The response of the epithelium of the tubes is slightly different. After a single estrogen injection there is little cell
division. If, however, estrogen treatment is continued, the mitotic maximum is attained 3 days after the beginning of treatment,
but suddenly falls off, even if treatment is continued. After
cessation of treatment, a new peak occurs 3 days after the last
injection. There is, thus, a marked difference in response to
estrogens by the surface epithelium and the epithelium of the tubes,
even though the origin of both is identical.

Progesterone inhibits the response of tubal epithelial cells
to estrogen stimulation, and given at the same time as an injection

of long-acting estrogen, stops the two mitotic peaks. If progesterone is given a day after a single injection of estrogens,
only the second mitotic peak is stopped. This suggests that during
the 24 hours between the two administrations, transcription of the
estrogen receptor takes place in the nucleus, and that DNA has been
synthesized.

In ovariectomized animals, a single injection of estrogens
or progesterone determines cellular multiplication. If daily injections of progesterone are given after an injection of estrogens, stromal cell division is maximal on day 3 of the progesterone
treatment. Moreover, if, a single dose of estrogens is injected
after progesterone injection, the stromal cells undergo an increased
rate of mitosis. This demonstrates that the stromal response is
strongly affected by previous progesterone and estrogen treatment.
Thus, it may be concluded that a preliminary injection of estrogen
and/or progesterone, or a combination of estrogen-progesterone
influences stromal mitoses (Finn and Martin, 1973; Finn and Porter,
1975).

Other Stromal Changes

The object of sequential estrogen-progesterone treatment is to
reproduce an endometrial state that closely resembles that of the
normal cycle. By studying the results of these treatments, it has
been possible to observe the synchronization of their sequential
and combined actions on the stroma. Two endometrial elements are
characteristic: the connective spines and the spiral arteries.

In combined estrogen-progestative treatments, both elements
are missing. In sequential treatment, the connective spines are
barely visible; sometimes they are rare but long, forming invaginations into the lumen, whereas the normal endometrium shows a
regular lacy appearance.

The spiral arteries are the result of transformation of the
capillaries by progesterone. They are characterized by the appearance of myofibrils in the pericytic cells, whose cytoplasm
widens and increases, forming an arteriolar wall without elastic
fibers (Ancla and de Brux, 1964). Stromal cells around the spiral
arteries have a decidual appearance. A very precise estrogen-
progesterone equilibrium is necessary for the formation of spiral
arteries and connective spines: an equilibrium that is quantitative
and also durationally controlled by progesterone. Differentiation
of the spiral arteries is realized by only a minute quantity of
progesterone in the presence of moderate or even slight, but continuous estrogen stimulation. Thus, in cycles with even slight
estrogen-progesterone imbalance, the presence of spiral arteries
pinpoints the date of the cycle and enables determination of the

degree of luteal defect by comparison with the secretion of the
epithelial cells. The connective spines necessitate correct balance of the two hormones; a predominance of estrogens induces
excessive proliferation. It is possible that other hormonal activities, (e.g., somatotrope) may also influence the appearance of
the spiral arteries and spines (Yen et al., 1970).

SUMMARY

    The ever intensified study of hormonal receptors permits a
better understanding of the histology of endometria subjected to
isolated or combined stimulation. Certain points, nevertheless,
still remain obscure, in particular, the different responses of
each constituent to the same hormonal stimuli. These differences
are probably due to various cellular differentiations, testifying
to their individual specific functions. The endometrium appears
to be a single entity, but each constituent has a special function,
the combination of which, if balanced, permits nidation.

REFERENCES

Ancla, M., and de Brux, J., 1964, Etude au microscope électronique
        des artérioles spiralées de l'endomètre humain, Ann. Anat.
        Pathol., 9:209.
Baulieu, E.E., Atger, M., Best-Belpomme, M., Corvol, P., Courvalin,
        J.C., Mester, J., Milgrom, E., Robel, P., Rochefort, H., and
        Decatalogne, D., 1975, Steroid hormone receptors, Vitam. Horm.,
        33:649.
Bouteille, M., Kalifat, S.R., and Delarue, J., 1967, Ultrastructural
        variations of nuclear bodies in human diseases, J. Ultrastruct.
        Res., 19:474.
Brenner, R.M., Resko, J.A., and West, N.B., 1974, Cyclic changes
        in oviduct morphology and residual cytoplasmic estradiol
        binding capacity induced by sequential estradiol-progesterone
        treatment of spayed rhesus monkeys, Endocrinology, 95:1094.
Clark, J.H., Hardin, J.W., Upchurch, S. and Eriksson, H., 1978a,
        Heterogenity of estrogen binding sites in the cytosol of rat
        uterus, J. Biol. Chem., 253:7630.
Clark, J.H., and Peck, E.J., 1979, Female sex steroids - Receptors
        and Functions, Springer Verlag, Berlin.
Clark, J.H., Peck, E.J., Hardin, J.W., and Erikkson, H., 1978b,
        Biology and pharmacology of estrogen receptors. Relation to
        physiological response, in: "Receptors and Hormone Action,"
        B.W. O'Malley, and L. Birnbaumer, ed., Vol. 2, Academic
        Press, New York, p. 1.
Dupuy-Coin, A.M., Kalifat, S.R., and Bouteille, M., 1972, Nuclear
        bodies as proteinaceous structures containing ribonucleo-
        proteins, J. Ultrastruct. Res., 38:174.

Finn, C.A., and Martin, L., 1973, Endocrine control of gland proliferation in mouse uterus, Biol. Reprod., 8:585.

Finn, C.A., and Porter, D.A., 1975, The action of ovarian hormones on the endometrium, in: "The Uterus Reproductive Biology Handbook," C.A. Finn and D.A. Porter, ed., Elek Science, London, p. 42.

Gorski, J., and Gannon, F., 1976, Current models of steroid hormone action: a critique, Annu. Rev. Physiol., 28:425.

Hertz, R., 1974, The estrogen-cancer hypothesis, Cancer, 38:534.

Hisaw, F.L., 1959, Comparative effectiveness of estrogen fluids, imbibition and growth of the rat's uterus, Endocrinology, 64:276.

Hsueh, A.J., Peck, E.J., and Clark, J.H., 1976, Control of uterine estrogen receptor levels by progesterone, Endocrinology, 98:438.

Huggins, C., and Jensen, E.V., 1955, The depression of estrone-induced uterine growth by phenolic estrogens with oxygenated functions at position 6 or 16: the impeded estrogens, J. Exp. Med., 102:335.

Jensen, E.V., Block, G.E., Smith, S., Kyser, K., and DeSombre, E.R., 1973, Estrogen receptor and hormone dependency, in: "Estrogen Target Tissue and Neoplasia," T.L. Dao, ed., University of Chicago Press, Chicago, p. 23.

Jensen, E.V., Mohla, S., Gorell, T.A., and DeSombre, E.R., 1974, The role of estrophilin in estrogen action, Vitam. Horm., 32:89.

Krishan, A., Uzman, B.G., Hedley-Whyte, E.T., 1967, Nuclear bodies: a component of cell nuclei in hamster tissues and human tumors, J. Ultrastruct. Res., 19:563.

Lacassagne, A., 1932, Apparition de cancers de la mammelle de la souris mâle à des injections de folliculine, C.R. Acad. Sci., Paris, 195:630.

Lawn, A.M., 1973, The ultrastructure of endometrium during the sexual cycle, Adv. Reprod. Physiol., 6:61.

Leavitt, W.W., Chen, T.J., Do, Y.S., Carlson, B.D., and Allen, T.C., 1978, Biology of progesterone receptor, in: "Receptors and Hormone Action," B.W. O'Malley and L. Birnbaumer, ed., Vol. 2, Academic Press, New York, p. 157.

Means, A.R., and O'Malley, B.W., 1971, Protein biosynthesis on chick oviduct polyribosomes. Regulation by progesterone, Biochemistry, 10:1570.

Milgrom, E., Perrot, M., Atger, M., and Baulieu, E.E., 1972, Progesterone in uterus and plasma. V. An assay of the progesterone cytosol receptor of the guinea pig uterus, Endocrinology, 90:1064.

Milgrom, E., Atger, M., and Baulieu, E.E., 1973, Mechanism regulating the concentration and conformation or progesterone receptors in uterus, J. Biol. Chem., 248:6366.

O'Malley, B.W., and Means, A.R., 1974, Female steroid hormones and
    target cell nuclei, Science, 183:610.
Pavlik, E.J., and Coulson, P.B., 1976, Modulation of estrogen recep-
    tors in four different target tissues: differentiation effects
    of estrogen vs progesterone, J. Steroid Biochem., 7:313.
Pollow, K., Schmidt-Gollwitzer, N., and Nevinny-Stockel, J., 1977,
    Progesterone receptors in normal human endometrium and endo-
    metrial carcinoma, in: "Progesterone Receptors in Normal and
    Neoplastic Tissues," W.L. McGuire, J.P. Raynaud, E.E. Baulieu,
    ed., Raven Press, New York, p. 313.
Warembourg, M., and Milgrom, E., 1977, Radioautography of the uterus
    and vagina after [^{3}H]progesterone injection into guinea pigs
    at various periods of the estrous cycle, Endocrinology, 100:
    175.
Wotiz, H.H., Shane, J.A., Vigerski, R., and Brenher, P.I., 1968,
    The regulatory role of estradiol in the proliferative action
    of estradiol, in: "Prognostic Factors in Breast Cancer,"
    A.P.M. Forrest, P.B. Kunkler, ed., Livingstone Press, Edin-
    burgh, p. 368.
Yamamo, K.R., and Alberts, B.M., 1976, Steroid receptors, elements
    for modulation of eukaryotic transcription, Ann. Res. Biochem.,
    45:721.
Yen, S.S.C., Vela, P., Rankin, J., and Littell, A.S., 1970, Hormonal
    relationships during the menstrual cycle, J. Am. Med. Assoc.,
    211:1513.

ESTRADIOL AND PROGESTERONE RECEPTORS IN HUMAN ENDOMETRIUM

Paul Robel, Rodrigue Mortel,* and Etienne Emile Baulieu

Unité de Recherches sur le Métabolisme Moléculaire
et la Physio-Pathologie des Stéroides
L'institut National de la Sante et de la Recherche
Médicale
(U 33 INSERM) and ER 125 CNRS
78 rue du Général Leclerc
94270 Bicetre, France

*Present Address:
Milton S. Hershey Medical Center
The Pennsylvania State University
Department of Obstetrics and Gynecology
Division of Gynecologic Oncology
Hershey, Pennsylvania

INTRODUCTION

In the last two decades, the development of saturation analysis methods has permitted specific and accurate measurement of
circulating sex steroid hormones and documentation of their cyclic
changes during the menstrual cycle. The large body of evidence
accumulated over the past ten years indicates that hormones interact with a receptor system before triggering cellular responses.
Hormone receptors were first identified and studied physiochem-
ically for steroids, particularly estradiol, in the rat uterus.
Once reliable measurements of hormone receptors became available
in animal models, correlations between receptor concentrations
and circulating levels of various hormones were attempted. It was
discovered that the concentration of receptor molecules is not
fixed and varies with the physiological state of animals. Variations in the concentration and subcellular distribution of receptors were observed during the estrus cycle and could be mimicked
in hormone-deprived animals by injecting estradiol and progesterone (Baulieu et al., 1975).

43

In immature rat uterus, estradiol receptor is found almost
exclusively in the soluble fraction of the cytoplasm commonly
called the cytosol. After injection of estradiol in vivo, estradiol-receptor complexes are translocated to the nuclear compartment of the cell. Considerable effort has been devoted to the
search for nuclear "acceptors", which may be the sites where characteristic changes in gene expression occur. However, it should
be stressed that the presence of the receptor in the nucleus does
not necessarily trigger hormone action. For example, antiestrogens
are capable of transferring the receptor into the nucleus, nevertheless no response may occur (Sutherland et al., 1977). Another
example is the finding in some target tissues, including human
endometrium, of readily available (apparently unoccupied) nuclear
receptor sites, which estradiol can reach directly. Therefore, it
appears that the formation of hormone-receptor complexes in the
cytoplasm is not mandatory for hormone action. In practical terms,
the cytosol to nuclear receptor ratio is worth considering, but
should be interpreted with caution.

The process of free and occupied receptor sites, both in
cytosol and nuclei, must be kept in mind. Because of the slow
dissociation rate of estradiol from receptor sites, techniques
have been designed to measure either available sites at low temperature or both available and occupied sites by "exchange" at
higher temperatures. The latter measurements are fraught with
difficulties because receptors are easily inactivated. Available
receptor sites may be either unoccupied, or occupied by low-affinity
natural or synthetic ligands. Finally, it should be emphasized
that techniques worked out to measure receptor sites occupied by
natural hormones (estradiol or progesterone) may have to be adjusted
in certain physiological or pharmacological conditions when receptor
sites are occupied by other ligands.

The presence of estradiol receptor in the human female reproductive tract was suggested by the selective retention of radioactive estradiol in the normal human uterus (Davis et al., 1963)
and endometrium (Brush et al., 1967; Evans and Hähnel, 1971). The
first quantitative approach for measurement of total receptor concentration in human endometrium applied the technique of endometrium
slice superfusion (Tseng and Gurpide 1972a). Wiest and Rao (1971)
are credited with the first unambiguous demonstration of progesterone
receptor in human endometrium. A systematic study of estradiol and
progesterone receptors in both cytoplasmic and nuclear fractions
was conducted by our group in normal women (Bayard et al., 1978).

METHODOLOGICAL PROBLEMS

General

It is usually impossible to obtain endometrial tissue from
women deprived of estrogens. However, when the endometrium is
exposed to endogenous hormones, some or all receptor sites may be
occupied by hormones and a large fraction of the hormone-receptor
complexes is found in the nuclear compartment. Therefore, interpretation of published results must take into account which fractions have been measured: (1) free cytoplasmic receptor sites;
(2) free and occupied cytoplasmic receptor sites; (3) free nuclear
receptor sites; (4) free and occupied nuclear receptor sites; or
(5) total cellular receptor sites. Indeed, the physiological
significance of the results will differ according to the type of
receptor site assayed. Unfortunately, many published reports are
difficult to interpret because the assay conditions are primarily
empirical, i.e., the assay measures only part of the available sites
when less than saturating concentrations of hormones are used.
Similarly, the assay determines part of the occupied sites when
exchange reactions are incomplete. In cases where receptor concentrations are determined after incubation of tissue slices with
radiolabeled hormones, it is clear that this method measures translocated cytosolic hormone-receptor complexes not endogenous nuclear
receptors. Such improper terminology coupled with improper methodology explain most discrepancies in the literature.

Labeling receptor sites, particularly occupied ones, with
tritiated hormones of high specific activity is based on equilibrium
and kinetic association and dissociation rate constants, and provides
the basis for the so-called "exchange" techniques: radioactive
hormone instead of nonradioactive hormone fills the receptor sites.
Besides the receptor, crude endometrial extracts contain other
hormone-binding components that may be divided into low-affinity
nonsaturable binding and high-affinity saturable binding components.
The latter are similar to plasma proteins that specifically bind
to natural steroid hormones. Consequently, binding of hormone to
the receptor must be differentiated from binding to both kinds of
nonreceptor binding components. A rational approach to such dis-
crimination is presently based on the physicochemical properties
of estradiol and progesterone receptors, i.e., affinity and specificity of the binding, heat stability, and other characteristics
related to size, electrophoretic mobility and sensitivity to various
reagents.

Endogenous Hormones in Human Endometrium

Proliferative endometrium and secretory endometrium contain
approximately 1 ng and 0.5 ng of estradiol/g of tissue respectively.

These amounts (Guerrero et al., 1975; Batra et al., 1977) produce concentrations of endogenous estradiol in cytosol of 0.5 nM or less, and are insignificant in most receptor assay procedures.

Progesterone concentration in the endometrium has been determined by several authors (Haukkamaa and Luukkainen, 1974; Bayard et al., 1975; Guerrero et al., 1975; Batra et al., 1977; Kreitmann et al., 1978). When recalculated in ng/g of tissue, rather widespread values were obtained. In the follicular phase, progesterone levels ranged from 2 to 9 ng/g of tissue, and values for the mid-secretory phase varied between 8.5 and 27 ng/g. Therefore, at least in the secretory endometrium, the concentration of endogenous progesterone causes a very significant isotopic dilution of added tracer. In fact, the concentration of progesterone receptor sites doubled when cytosol was stripped of endogenous progesterone using the dextran-coated charcoal procedure (Haukkamaa, 1974).

Plasma Proteins in Endometrial Cytosol

As initially demonstrated in the rat uterus (Milgrom and Baulieu, 1970), a major difficulty encountered in developing an assay to measure progesterone receptor is the presence of a corticosteroid-binding globulin (CBG)-like component in uterine cytosol, in spite of thorough washing of tissues before homogenization. In a report by Verma and Laumas (1973), the only observed binding component showed the physicochemical properties and binding specificity of CBG. In other reports, a combination of physical methods such as gradient ultracentrifugation, polyacrylamide gel electrophoresis and appropriate binding competition experiments (Young and Cleary, 1974; Kreitmann et al., 1978) indicated the simultaneous presence of progesterone receptor and CBG.

An absolute requirement in setting up the assay is to demonstrate that progesterone can only be displaced from binding sites by progestagens and not by cortisol. A convenient approach is to determine the binding of [^{3}H]progesterone in the presence of competitors and a 100-to 1,000 fold excess of nonradioactive cortisol. This technique prevents the binding of progesterone to CBG (Bayard et al., 1978) by saturation of CBG-binding sites with nonradioactive cortisol. Kreitmann et al. (1978) have measured the concentration of CBG in human endometrium and have observed large variations among samples, which were apparently unrelated to the phase of the menstrual cycle. The reported values averaged 25 pmol/g, which represent approximately 5% of the plasma values (about 500 nM).

In endometrial cytosol, the concentration of CBG is generally severalfold greater than that of progesterone receptor (Kreitmann et al., 1978). Since the affinity of progesterone for CBG and receptor is similar, it is quite conceivable that most saturable binding sites in endometrial samples with low receptor content may be due to contamination by plasma CBG. No enrichment of endometrial cytosol in CBG versus other plasma proteins has been observed (Kreitmann et al., 1978).

The sex hormone binding plasma protein (SBP) is one of the plasma proteins that contaminates human endometrium. However, its concentration in nonpregnant female plasma is ten times lower than that of CBG, approximately 60 nM (Heyns and De Moor, 1971). Similarly, SBP affinity for estradiol (K_{Deq} 1.3 nM at 4° C; Mercier-Bodard et al., 1970), is much lower than that of estradiol for the receptor. Consequently, the binding of estradiol to SBP in cytosol preparations can generally be considered negligible (Bayard et al., 1978).

Metabolism of Ligands

Following the work of Tseng and Gurpide (1972a,b), estradiol dehydrogenase activity has been well demonstrated in human endometrium and its changes have been correlated with hormonal status. This enzyme converts estradiol to estrone and is located mainly in the glandular epithelium of secretory endometrium (Scublinski et al., 1976), particularly in endoplasmic reticulum and mitochondrial subcellular fractions (Pollow et al., 1975b). NAD is the preferred cofactor and the K_m for estradiol is 3.3 μM at 40° C and pH 9.5. The oxidation of estradiol to estrone by estradiol dehydrogenase is greatly enhanced in secretory endometrium, as shown by Tseng and Gurpide (1972b, 1974) as well as by Pollow et al. (1975b). In addition, the activity of this enzyme can be increased by progestagens either *in vivo* or *in vitro* (Tseng and Gurpide, 1975a).

Progesterone is also metabolized by human endometrium, as reported by Sweat and Bryson (1970). According to these authors, the major metabolites were 5α-pregnane-3-20-dione (pregnanedione), an undefined dihydroxy-compound, and 6β-hydroxyprogesterone. However, when NADPH was added, the major products were pregnanedione and 20α-OH-pregn-4-ene-3-one (20α-OH progesterone) (Collins and Jewkes, 1974). Pollow et al. (1975c) demonstrated 5α-reductase, 5β-reductase and 20-hydroxysteroid dehydrogenase activity. The latter has been reported to be more active during the secretory phase of the menstrual cycle. However, these enzyme activities do not preclude the use of corresponding natural hormones in receptor measurements because no significant ligand metabolism was observed under the assay conditions (Hähnel, 1971; Bayard et al., 1978). This lack of ligand metabolism is mainly due to the localization of most enzyme activities in the particulate fractions of the cytoplasm and to the relatively large K_m values of enzyme compared to hormone concentrations needed for receptor assays.

However, the in vivo metabolism of estradiol and progesterone in
human endometrium partly explains the dissimilarities between
plasma and tissue concentrations of both hormones.

Main Characteristics of Reliable Assays for Estradiol and Progesterone Receptors in Human Endometrium

General. Pathophysiological changes in receptor concentrations
cannot be evaluated without a thorough validation of the specificity of the technique adopted. It is imperative that the binding
assay is specific for estradiol and/or progesterone receptor and
that it gives an accurate evaluation of binding constants. It
must be realized that no evidence exists for significant affinity
changes of receptors with physiological states and, therefore,
the only parameter that is subjected to variation is the concentration of receptor sites. It is equally important to be mindful
that human endometrium is generally exposed to endogenous ovarian
hormones, and that receptor sites can be expected to be filled or
unfilled with hormones and located both in the cytoplasm and in
the nuclear fractions. It should also be recalled that, since
estradiol dissociates very slowly from receptor sites at low
temperature, only unfilled sites will be measured under these
conditions, whereas the sum of filled and unfilled sites will be
determined when an exchange technique is used (Anderson et al.,
1972). Conversely, since the dissociation of progesterone-receptor
complexes is very rapid, some exchange will necessarily occur
regardless of the assay conditions. Therefore, in the case of
progesterone receptors, only the sum of unfilled and occupied
receptor sites can be rigorously measured.

Choice of ligand. The criteria for selection of the best
radioactive ligand are: high affinity, slow dissociation rate,
strict binding specificity and nonmetabolizability. In the case
of estrogen receptors, [3H]estradiol has generally been utilized,
since it fulfills roughly all of these conditions. Assay conditions
have been described where no metabolism occurs and where binding to
SBP and to androgen receptor (present in very minute amounts,
(unpublished data), has been prevented by adding 20 nM dihydro-
testosterone to the incubation buffers (Bayard et al., 1978).
Recently, the use of [3H]moxestrol has been proposed by Raynaud
et al. (1978), because it binds minimally to SBP, androgen receptor,
and nonsaturable low-affinity proteins.

In the case of progesterone receptors, the use of [3H]proges-
terone is acceptable, providing that the rapid dissociation of
progesterone-receptor complexes is controlled by addition of glycerol and that binding of the hormone to CBG is prevented by addition
of excess nonradioactive cortisol to the incubation buffers (Feil
et al., 1972; Milgrom et al., 1972; Bayard et al., 1978). As an
alternative, a synthetic progestagen, such as [3H]R5020, which

does not bind to CBG, can be used (Raynaud, 1977); however, this
compound binds rather strongly to nonsaturable components (Seematter et al., 1978), and also displays relatively high affinity
for glucocorticosteroid receptors. The latter difficulty can be
circumvented by the addition of cortisol to assay buffers. Another
synthetic progestin, [^{3}H]ORG 2058, seems to have properties similar
to those of R5020 (Jänne et al., 1976).

    <u>Choice of binding assay</u>. Few systematic attempts have been
made to define the most efficient and accurate techniques for
measuring receptors. In general, low ionic-strength buffers
containing EDTA, SH-reducing agents and glycerol are used at a
slightly alkaline pH (7.8 to 8), because they are thought to protect
cytosol receptors. The dextran-coated charcoal adsorption technique
has been utilized in most reports for binding assays (Wiest and
Rao, 1971; Krishnan et al., 1973; Crocker et al., 1974; Evans et
al., 1974; Rao et al., 1974; Young and Cleary, 1974; MacLaughlin
and Richardson, 1976; Bayard et al., 1978), and, likewise, gel
filtration (Trams et al., 1973; Makler and Eisenfeld, 1974), has
been employed. Several authors (Verma and Laumas, 1973; Philibert
and Raynaud, 1974) have reported that the use of equilibrium dialysis
results in large losses of binding sites. More selective procedures
such as ammonium sulfate or protamine sulfate precipitation have
not been currently adopted.

    The measurement of binding site concentration can be performed
either by constructing a Scatchard plot or by performing a single
point analysis at a saturating concentration of ligand. The former
approach allows K_{Deq} determination, thus confirming the specificity
of the measured binding. However, in the presence of endogenous
hormone, once isotopic equilibrium has been reached, the plot
provides only an acceptable estimate of receptor site concentration,
and an underestimation of the affinity constant. The single point
analysis is determined by labeling the sites with a saturating concentration of radioactive hormone. It requires smaller amounts of
material, and under carefully controlled conditions, gives results
similar to the multipoint analysis necessary for constructing a
Scatchard plot (Haukkamaa and Luukkainen, 1974).

    Regardless of the method employed, binding to nonsaturable
proteins must be subtracted by parallel incubation(s) with the
same concentration(s) of radioactive hormone plus a large excess
of nonradioactive hormone. On occasion, nonsaturable binding has
been evaluated after inactivation of the receptor by heat or SH
reagents. Binding to specific plasma proteins or possibly to other
receptors must be prevented, as previously indicated, by the use
of appropriate radioactive ligands and/or by the addition of
appropriate nonradioactive competitors.

In the case of progesterone receptor, endogenous progesterone, depending on its concentration, may interfere with the binding assay. This difficulty can be prevented by exposing the cytosol to dextran-coated charcoal at 4^0 C for $\leq$ 30 min., a procedure that removes most endogenous progesterone prior to incubation (Young and Cleary, 1974). Similarly, the concentration of endogenous progesterone can be measured by radioimmunoassay and the result can be used to correct for the specific activity of the added tracers (Bayard et al., 1978). If this correction is not made, the concentration of progesterone receptor sites can be underestimated by as much as 33 to 50% of the actual values. This is particularly true for progesterone receptor measurement in gestational endometrium.

Practical constraints. For clinical investigation purposes, measurement of receptor concentration is most often performed on endometrial samples obtained by biopsy. The weight of such samples rarely exceeds 200 mg, and part must be kept for histological examination. Therefore, it becomes advantageous to use assay techniques that allow measurement of both estradiol and progesterone receptor concentrations in cytosol prepared from samples equal to or greater than 50 mg. Recently Levy et al. introduced a glass fiber filter exchange assay that permits the measurement of nuclear receptors on the same amount of tissue (unpublished data).

The exact location of the biopsy material taken from the uterine cavity is often unknown, and interpretation of receptor assays is hindered by the fact that receptors are not evenly distributed throughout the endometrial lining. This explains the numerous conflicting results that have been published. Robertson et al. (1971) and Tsibris et al. (1978) reported that the concentration of receptor in endometrium showed a progressive decrease throughout the length of the organ from the fundus to the cervix, whereas Lunan and Green (1975) reported more receptors in the body than in the fundus. Brush et al. (1967) indicated that in some cases the uptake of estradiol by different regions of the endometrium varied considerably. Bayard et al. (1978) also reported large differences in receptor concentration between biopsies of the fundus and the body of the uterus; however, they concluded that no systematic trend was discernible, and that the ratio of progesterone to estradiol receptors remained practically constant. In any case, a compromise can be reached by routinely taking biopsies from the midregion of the endometrium.

Endometrial samples kept in isotonic saline or homogenization buffer at 0 to 4^0 C must be processed relatively rapidly (within 1 hr) to prevent inactivation of receptor sites. Receptors are also inactivated when endometrial curettings are kept in liquid nitrogen. However, cytosol and nuclear estrogen and progesterone receptors remain stable when endometrial samples are immersed in a preserving medium and then frozen in liquid nitrogen (Bayard et al., 1978). Likewise, Koenders et al. (1978) reported no loss of cytosol receptor sites if samples are frozen in liquid nitrogen, lyophilized in glass vials stoppered under vacuum conditions and stored at 4°C. Both techniques allow shipment of endometrial biopsies from clinical departments to remote biochemical laboratories and permit convenient and simultaneous receptor measurement in small series of samples.

ESTRADIOL AND PROGESTERONE RECEPTORS IN NORMAL MENSTRUAL CYCLE

The dating of normal endometrium is usually based on a combination of several criteria: day of cycle (when the regularity of cycles is known), basal body temperature, histological evaluation and serial measurements of plasma LH, estradiol and progesterone. At best, the dating of the biopsy can be made with $\pm$ 1 day precision. In most reports dealing with receptor measurement, only broad terms such as "proliferative" and/or "secretory" endometrium are used. In the few publications where receptors were measured on a daily basis, the criteria for dating were usually not well defined. In our opinion, valuable information can be obtained by separate investigation of the four periods of the menstrual cycle: (1) early proliferative phase; (2) late proliferative phase, when the preovulatory plasma estradiol surge occurs; (3) early secretory phase, extending from the day of ovulation to the day of implantation; and (4) late secretory phase.

Estradiol Receptor

Retention of [³H]estradiol in whole tissue. The first indication of variable retention of [³H]estradiol during the menstrual cycle came from in vivo experiments. Two hours after intravenous injection of radioactive hormone, the concentration of radioactivity was higher in proliferative than in secretory endometrium (Brush et al., 1967). This was confirmed by in vitro incubation of tissue slices (Evans and Hähnel, 1971; Trams et al., 1973). Tseng and Gurpide (1972a) measured the amount of estradiol bound tightly to nuclei after in vitro superfusion or incubation of endometrial slices. Under carefully controlled conditions, the amount of nuclear estradiol-receptor complexes was found to be 3.1 pmol/mg DNA in proliferative, 1.6 pmol/mg in early secretory, 0.6 pmol/mg in midsecretory and 0.5 pmol/mg in the late secretory endometrium (Gurpide et al., 1976; Tseng et al., 1977). In a similar approach, Crocker et al. (1974) observed that the nuclear uptake of estradiol reached a peak in the late proliferative phase.

Cytosol receptor. In most reports, the adopted methodology allowed measurement of unfilled receptor sites, and failed to account for the portion occupied by hormone. Under these conditions, receptor concentration was found to be higher in proliferative than

in secretory endometrium (Trams et al., 1973; Crocker et al.,
1974; Evans et al., 1974; Pollow et al., 1975a; Schmidt-Gollwitzer
et al., 1978). In one published article, a maximum of 2.3 pmol/mg
of cytosol protein was measured in the late proliferative phase
(Crocker et al., 1974). All other publications report a continuous
decrease in primarily unfilled receptor sites from early proliferative until late secretory phase. This trend is undoubtedly due
to a progressive increase in occupied receptor sites because when
the incubation temperature was raised to 25° C, the concentration
of receptors increased more than two fold (Evans et al., 1974). In
addition, an inverse relationship was observed between the apparent
K_D and the apparent concentration of receptor, a situation typical
of measurements performed in the presence of endogenous hormone
and nonsteady state conditions (Pollow et al., 1976).

        Several authors have determined the sum of unfilled and
occupied receptor sites after exchange (Robertson et al., 1971;
Bayard et al., 1978; Sanborn et al., 1978; Levy et al., 1980a).
All agree that an increased amount of estradiol receptor appears
at midcycle and report mean values between 1.2 and 3.5 pmol/mg DNA.

        Nuclear receptor. Several investigators have measured spe-
cific nuclear binding after incubation of tissue slices with estradiol. However, it should be recognized that this technique measures
nuclearly translocated hormone receptor complexes not endogenous
nuclear receptors. An exchange assay of nuclear receptors in human
endometrium has been reported (Bayard et al., 1978). The concentration of nuclear receptor sites doubled between the early and late
proliferative phase, and almost equalled the concentration of the
cytosolic sites. Then, during the secretory phase, the nuclear
receptor level decreased, but at a slower rate than the cytoplasmic
receptor level. Recently, it has been shown that endometrial
nuclei may contain a significant proportion of receptor sites that
can be labeled with [3H]estradiol at 0° C, and presumably are
unoccupied by endogenous hormones (Levy et al., 1980b).

Progesterone Receptor

        Cytosol receptor. Because of the rapid dissociation of endo-
genous progesterone-receptor complexes during assay conditions,
both unfilled and occupied receptor sites have been measured by
most authors, often without considering isotopic dilution due to
endogenous progesterone. Binding to CBG has been eliminated
either by the addition of cortisol to incubations with progesterone
or by the use of synthetic progestagens. In some early publications,
very large values were reported (Haukkamaa and Luukkainen, 1974;
Rao et al., 1974) without significant changes throughout the cycle.
However, there is general agreement that a marked increase in
receptor concentration occurs in the late proliferative, or midcycle period (MacLaughlin and Richardson, 1976; Bayard et al., 1978;

Sanborn et al., 1978; Syrjälä et al., 1978; Levy et al., 1980a).
Although early papers published by Pollow's group (Pollow et al.,
1975a, 1976) report very low amounts of progesterone receptor
in the proliferative phase, their recent reports agree that a midcycle rise does occur (Schmidt-Gollwitzer et al., 1978).

There is some divergence of opinion concerning the absolute
concentration of progesterone receptor. In the late proliferative
phase, the mean values reported and expressed in pmol/mg were:
0.7 (MacLaughlin and Richardson, 1976), 2.8 (Levy et al., 1980a),
12 (Syrjälä et al., 1978), 23 (Sanborn et al., 1978), and 30
(Schmidt-Gollwitzer et al., 1978). However, the levels of progesterone receptor are always severalfold higher than those of estradiol receptor during the same phase of the menstrual cycle. The
highest values were obtained after use of synthetic progestins.
The reasons for this discrepancy are not quite clear. However, it
has been shown that in addition to binding to progesterone receptor, R5020 (the most widely used synthetic progestagen) has considerable affinity for glucocorticoid receptors, and appears to bind
to serum and tissue proteins with greater avidity than progesterone
(Lippman et al., 1977; Seematter et al., 1978; Powell et al., 1979).

Nuclear receptor. Only our group has published results concerning the nuclear receptor (Bayard et al., 1978; Levy et al.,
1980a) and our values have been consistently lower than those
observed for the cytosol receptor. A definite increase occurred
in the early secretory phase, reaching 0.6 pmol/mg DNA (2,500
sites/cell).

Hormonal Correlations of Estradiol and Progesterone Receptors

The simultaneous increases in estradiol and progesterone
receptors at midcycle follow the plasma estradiol surge during
the late proliferative phase and strongly suggest a positive relationship between estradiol and endometrial receptors. Indeed,
during the proliferative phase, estradiol blood levels correlate
positively with total progesterone and estradiol receptor sites
(Levy et al., 1980a) and also with cytoplasmic progesterone receptor sites (Schmidt-Gollwitzer et al., 1978).

A similar correlation was also reported between progesterone
receptor and cytosol estradiol in human myometrium (Kontula, 1975).
However, the concentration of unoccupied estradiol receptor sites
correlated inversely with the concentration of estradiol in the
blood (Trams et al., 1973; Schmidt-Gollwitzer et al., 1978). The
inductive action of estradiol was confirmed in estrogen-treated
postmenopausal women by the appearance of progesterone receptors
in endometrial cytosol (Jänne et al., 1975), and by the elevation
of myometrial progesterone receptor to values approaching those
reported during the proliferative phase (Illingworth et al., 1975).

    The large decrease in both estradiol and progesterone receptors
during the secretory phase occurs when plasma and endometrial progesterone concentrations are increased. Indeed, a negative correlation has been reported between plasma progesterone and cytosol
estradiol and progesterone receptors (Jänne et al., 1975; Schmidt-
Gollwitzer et al., 1978). However, in one report where statistical
evaluation was performed, this inverse relationship was not significant (Levy et al., 1980a). Tseng and Gurpide (1975b) have demonstrated that progesterone and synthetic progestins reduce the level
of estradiol receptor in human endometrium.

    Another important feature of progesterone action is the large
increase in microsomal 17β-hydroxysteroid dehydrogenase activity in
secretory endometrium reported by Tseng and Gurpide (1974) as well
as Pollow et al. (1975b). This effect explains the relatively
low concentration of estradiol in endometrium during the secretory
phase (Schmidt-Gollwitzer et al., 1978) and may account partly for
the antiestrogenic characteristics of progestins (Gurpide et al.,
1977). Human endometrial estradiol dehydrogenase showed a severalfold increase under the influence of progestins, either in vivo
or after in vitro incubation (Tseng and Gurpide 1975a; Pollow et
al., 1978). Since progesterone increases estradiol dehydrogenase
activity and decreases progesterone receptor levels, an inverse
relationship between both parameters was predicted and indeed was
recently reported by Levy et al. (1980a), in contrast to an earlier
article by Schmidt-Gollwitzer et al. (1978).

GENERAL CONCLUSIONS

    The changes in estradiol and progesterone receptors in human
endometrium throughout the normal menstrual cycle are in keeping
with findings previously described in experimental animals. The
increase in estradiol receptor during the preovulatory period is
related to the plasma estradiol surge, leading to synthesis of
more receptor in the cytoplasm and to nuclear translocation of the
hormone-receptor complexes (Fig. 1). At the same time, cytoplasmic
progesterone receptor increases as a result of the inductive effect
of estradiol (Milgrom et al., 1972, 1973; Brenner et al., 1974).

    The postovulatory decrease in the estradiol receptor might
be related to progesterone effects on conversion of estradiol to
estrone and on the level of the estradiol receptor (Mester et al.,
1974; Hsueh et al., 1975). Consequently, there is a decrease in
estrogen-dependent synthesis of progesterone receptor. In addition,
it has been observed that progesterone directly "inactivates" its
own receptor (Milgrom et al., 1973; Brenner et al., 1974; Tseng et
al., 1977). However, only cytoplasmic receptor sites decrease
immediately after ovulation, whereas the nuclear receptor sites of
both receptors do not. Instead, the nuclear progesterone receptor

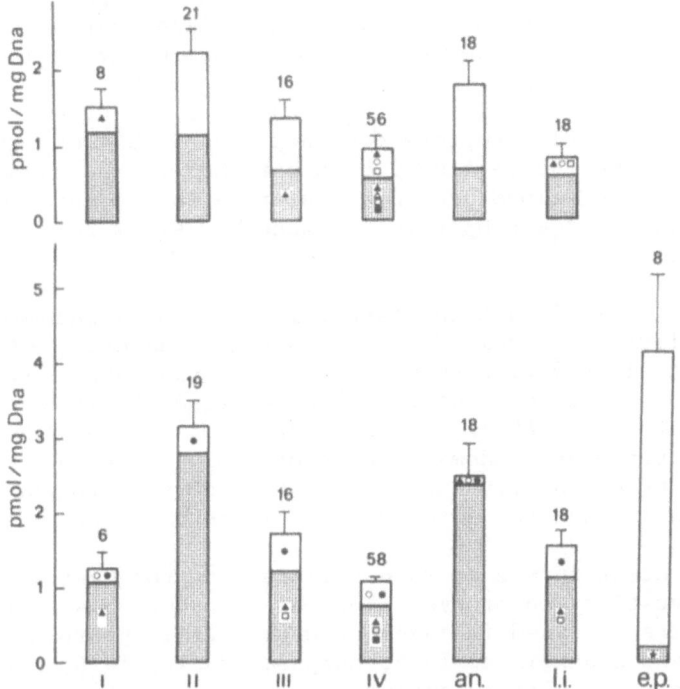

Figure 1.   Cytoplasmic and nuclear estradiol (upper panel) and
             progesterone (lower panel) receptors in normal and
             abnormal menstrual cycles and in early pregnancy.

             The rectangles represent the sum of cytoplasmic and
             nuclear receptor sites ± s.e.m.

             Shaded areas = cytoplasmic receptors; clear areas =
             nuclear receptors.

             Abbreviations: I, early proliferative phase; II, late
             proliferative phase; III, early secretory phase; IV,
             late secretory phase; an., anovulatory cycles; l.i.,
             luteal insufficiency; and e.p., early pregnancy. The
             number of samples investigated are indicated above each
             column. Symbols indicate the probability of considered
             concentration being significantly lower (at the 5%
             level) than the corresponding concentration of:
             I - Δ ; II - ▲ : III - ○ ; □ - an; ■ - l.i.; ● - e.p.;
             ★ - I, II, III, an. and l.i.

concentration increases, probably as a consequence of sustained nuclear transfer of hormone-receptor complexes (Bayard et al., 1978; Levy et al., 1980a). This observation stresses the critical importance of measuring both filled and unfilled cytosolic and nuclear receptor sites whenever receptor physiology is investigated.

During the menstrual cycle the concentration of nuclear progesterone and estradiol receptors does not exceed that of cytoplasmic receptors because the levels of both hormones in the endometrium are far below the values needed to saturate the receptor sites. Consequently, a large proportion of receptor sites remain unoccupied in the cytoplasm.

However, this is not the case for gestational endometrium in which very little cytosol (MacLaughlin and Richardson, 1976; Kreitmann et al., 1978; Levy et al., 1980a) and large concentrations of nuclear progesterone receptor were reported (Kreitmann et al. 1978; Levy et al., 1980a). The concentration of endogenous progesterone in gestational endometrium is markedly elevated and might explain the predominantly nuclear location of the progesterone receptors in the decidua after 8 to 12 weeks of pregnancy.

The interest in sex steroid hormone receptor measurement for clinical investigation of gynecological disorders has been heightened by the results obtained in cases of endometrial hyperplasia (Haukkamaa and Luukkainen, 1974; Gurpide et al., 1976; Syrjälä et al., 1978), anovulatory cycles and luteal insufficiency (Levy et al., 1980a). The high risk of endometrial hyperplasia and adenocarcinoma in patients with anovulatory cycles (Gusberg, 1976) may prove to be related to unopposed prolonged estradiol secretion, as it creates sustained high concentrations of nuclear estradiol-receptor complexes and a state of estrogen hyper-receptivity.

Recent reviews dealing with the application of progesterone and estradiol receptor measurements in human endometrial adeno-carcinoma have been published (McGuire et al., 1977; Brush et al., 1978; Richardson and MacLaughlin, 1978). Knowledge concerning the hormonal control of these receptors has permitted evaluation of biochemical changes in endometrial adenocarcinoma samples after in vivo challenge with progestagen, estrogen, or antiestrogen (Robel et al., 1978). Preliminary results are encouraging, and, if confirmed, a more rational program could be designed for treatment of patients with advanced or metastatic endometrial cancer.

REFERENCES

Anderson, J., Clark, J.H., and Peck, E.J., 1972, Oestrogen and
        nuclear binding sites. Determination of specific sites by
        [^{3}H]oestradiol exchange, Biochem. J., 126: 561.

Batra, S., Grundsell, H., and Sjöberg, N.-O., 1977, Estradiol-17β and progesterone concentrations in human endometrium during the menstrual cycle, Contraception, 16:217.

Baulieu, E.E., Atger, M., Best-Belpomme, M., Corvol, P., Courvalin, J.C., Mester, J., Milgrom, E., Robel, P., Rochefort, H., and De Catalogne, D., 1975, Steroid hormone receptors, Vitam. Horm., 33:649.

Bayard, F., Louvet, J.P., Monrozies, M., Boulard, A., and Pontonnier, G., 1975, Endometrial progesterone concentrations during the menstrual cycle, J. Endocrinol. Metab., 41:412.

Bayard, F., Damilano, S., Robel, P., and Baulieu, E.E., 1978, Cyto-plasmic and nuclear estradiol and progesterone receptors in human endometrium, J. Clin. Endocrinol. Metab., 46:635.

Brenner, R.M., Resko, J.A., and West, N.B., 1974, Cyclic changes in ovaductal morphology and residual cytoplasmic estradiol binding capacity induced by sequential estradiol-progesterone treatment of spayed rhesus monkeys, Endocrinology, 95:1094.

Brush, M.G., Taylor, R.W., and King, R.J.B., 1967, The uptake of [6,7-^{3}H]oestradiol by the normal human female reproductive tract, J. Endocrinol., 39:599.

Brush, M.G., King, R.J.B., and Taylor, R.W., 1978, "Endometrial Cancer," Baillière Tindall, London.

Collins, J.A., and Jewkes, D.M., 1974, Progesterone metabolism by proliferative and secretory human endometrium, Am. J. Obstet. Gynecol., 118:179.

Crocker, S.G., Milton, P.J.D., and King, R.J.B., 1974, Uptake of [6,7^3-H]oestradiol-17β by normal and abnormal human endome-trium, J. Endocrinol., 62:145.

Davis, M.E., Wiener, M., Jacobson, H.I., and Jensen, E.V., 1963, Estradiol metabolism in pregnant and nonpregnant women, Am. J. Obstet. Gynecol., 87:979.

Evans, L.H., and Hähnel, R., 1971, Oestrogen receptors in human uterine tissue, J. Endocrinol., 50:209.

Evans, L.H., Martin, J.D., and Hahnel, R., 1974, Estrogen receptor concentration in normal and pathological human uterine tissues, J. Clin. Endocrinol. Metab., 38:23.

Feil, P.D., Glasser, S.R., Toft, D.O., and O'Malley, B.W., 1972, Progesterone binding in the mouse and rat uterus, Endocrin-ology, 91:738.

Guerrero, R., Landgren, B.-M., Monteil, R., Cekan, Z., and Diczfalusy, E., 1975, Unconjugated steroids in the human endometrium, Contraception, 11:169.

Gurpide, E., Gusberg, S., and Tseng, L., 1976, Estradiol binding and metabolism in human endometrial hyperplasia and adeno-carcinoma, J. Steroid Biochem., 7:891.

Gurpide, E., Tseng, L., and Gusberg, S.B., 1977, Estrogen metabolism in normal and neoplastic endometrium, Am. J. Obstet. Gynecol., 129:809.

Gusberg, S., 1976, The individual at high risk for endometrial
        carcinoma, Am. J. Obstet. Gynecol., 126:535.
Hahnel, R., 1971, Properties of the estrogen receptor in the
        soluble fraction of human uterus, Steroids, 17:105.
Haukkamaa, M., 1974, Binding of progesterone by rat myometrium
        during pregnancy and by human myometrium in late pregnancy,
        J. Steroid Biochem., 5:73.
Haukkamaa, M., and Luukkainen, T., 1974, The cytoplasmic proges-
        terone receptor of human endometrium during the menstrual
        cycle, J. Steroid·Biochem, 5:447.
Heyns, W., and De Moor, P., 1971, The binding of 17β-hydroxy-5α-
        androstan-3-one to the steroid-binding β-globulin in human
        plasma, as studied by means of ammonium sulfate precipitation,
        Steroids, 18:709.
Hsueh, A.J., Peck, E.J., and Clark, J.H., 1975, Progesterone
        antagonism of the oestrogen receptor and oestrogen-induced
        uterine growth, Nature, 254:337.
Illingworth, D.V., Wood, G.P., Flickinger, G.L., and Mikhail, G.,
        1975, Progesterone receptor of the human myometrium, J. Clin.
        Endocrinol. Metab., 40:1001.
Jänne, O., Kontula, K., Luukkainen, T., and Vihko, R., 1975, Oes-
        trogen-induced progesterone receptor in human uterus, J.
        Steroid Biochem., 6:501.
Jänne, O., Kontula, K., and Vihko, R., 1976, Progestin receptors
        in human tissues: concentrations and binding kinetics, J.
        Steroid Biochem., 7:1061.
Koenders, A.J., Geurts-Moespot, J., Kho, K.H., and Benraad, Th.J.,
        1978, Estradiol and progesterone receptor activities in
        stored lyophilised target tissue, J. Steroid Biochem., 9:947.
Kontula, K., 1975, Progesterone-binding protein in human myometrium.
        Binding site concentration in relation to endogenous proges-
        terine and estradiol-17β levels, J. Steroid Biochem., 7:1555.
Kreitmann, B., Derache, B., and Bayard, F., 1978, Measurement of
        the corticosteroid-binding globulin, progesterone, and
        progesterone "receptor" content in the human endometrium,
        J. Clin. Endocrinol. Metab., 47:350.
Krishnan, A.R., Hingorani, V., and Laumas, K.R., 1973, Binding of
        ^{3}H-oestradiol with receptors in the human endometrium and
        myometrium, Acta Endocrinol., 74:756.
Levy, C., Robel, P., Gautray, J.P., de Brux, J., Verma, U., 1980a,
        Estradiol and progesterone receptors in human endometrium,
        normal and abnormal menstrual cycles and early pregnancy,
        Am. J. Obstet. Gynecol., 136:646.
Levy, C., Mortel, R., Eychenne, B., Robel, P., and Baulieu, E.E.,
        1980, Unoccupied nuclear oestradiol-receptor sites in normal
        human endometrium, Biochem. J., 185:733.
Lippman, M., Huff, K., Bolan, G., and Neifeld, J.P., 1977, Inter-
        actions of R5020 with progesterone and glucorticoid receptors
        in breast cancer and peripheral blood lymphocytes in vitro,

in: "Progesterone Receptors in Normal and Neoplastic Tissues," W.L. McGuire, J.P. Raynaud, and E.E. Baulieu, ed., Raven Press, New York, p. 193.

Lunan, C.B., and Green, B., 1975, Oestradiol-17β uptake in vitro into the nuclei of endometrium from different regions of the human uterus, Acta Endocrinol., 78:353.

McGuire, W.L., Raynaud, J.P., and Baulieu, E.E., 1977, Progesterone receptors: introduction and overview, in: "Progesterone Receptors in Normal and Neoplastic Tissues," W.L. McGuire, J.P. Raynaud, and E.E. Baulieu, ed., Raven Press, New York, p. 1.

MacLaughlin, D.T., and Richardson, G.S., 1976, Progesterone binding by normal and abnormal human endometrium, J. Clin. Endocrinol. Metab., 42:667.

Makler, A., and Eisenfeld, A.J., 1974, In vitro binding of ^{3}H-estradiol to macromolecules from the human endometrium, J. Clin. Endocrinol. Metab., 38:628.

Mercier-Bodard, C., Alfsen, A., and Baulieu, E.E., 1970, Sex steroid binding plasma protein (SBP), Acta Endocrinol., 147:204.

Mester, I., Martel, D., Psychoyos, A., and Baulieu, E.E., Hormonal control of oestrogen receptor in uterus and receptivity for ovoimplantation in the rat, Nature, 250:776.

Milgrom, E., and Baulieu, E.E., 1970, Progesterone in uterus and plasma. I. Binding in rat uterus 105,000g supernatant, Endocrinology, 87:276.

Milgrom, E., Atger, M., Perrot, M., and Baulieu, E.E., 1972, Progesterone in uterus plasma. VI. Uterine progesterone receptors during the estrus cycle and implantation in the guinea pig, Endocrinology, 90:1071.

Milgrom, E., Thi, M.L., Atger, M., and Baulieu, E.E., 1973, Mechanisms regulating the concentration and the conformation of progesterone receptor(s) in the uterus, J. Biol. Chem., 248:6366.

Philibert, D., and Raynaud, J.P., 1974, Binding of progesterone and R 5020, a highly potent progestin, to human endometrium and myometrium, Contraception, 10:457.

Pollow, K., Lübbert, H., Boquoi, E., Kreutzer, G., and Pollow, B., 1975a, Characterization and comparison of receptors of 17β-estradiol and progesterone in human proliferative and endometrial carcinoma, Endocrinology, 96:319.

Pollow, K., Lübbert, H., Boquoi, E., Kreutzer, G., Jeske, R., and Pollow, B., 1975b, Studies on 17β-hydroxysteroid dehydrogenase in human endometrium and endometrial carcinoma. I. Subcellular distribution and variations of specific enzyme activity, Acta Endocrinol., 79:134.

Pollow, K., Lübbert, H., Boquoi, E., and Pollow, B., 1975c, Progesterone metabolism in normal endometrium during the menstrual cycle in endometrial carcinoma, J. Clin. Endocrinol. Metab., 41:729.

Pollow, K., Boquoi, E., Schmidt-Gollwitzer, M., and Pollow, B.,
    1976, The nuclear estradiol and progesterone receptors of
    human endometrium and endometrial carcinoma, J. Mol. Med.,
    1:325.
Pollow, K., Schmidt-Gollwitzer, M., Boquoi, E., and Pollow, B.,
    1978, Influence of estrogens and gestagens on 17β-hydroxy-
    steroid dehydrogenase in human endometrium and endometrial
    carcinoma, J. Mol. Med., 3:81.
Powell, B., Garola, R.E., Chamness, G.C., and McGuire, W.L., 1979,
    Measurement of progesterone receptor in human breast cancer
    biopsies, Cancer Res., 39:1678.
Rao, B.R., Wiest, W.G., and Allen, W.M., 1974, Progesterone "recep-
    tor" in human endometrium, Endocrinology, 95:1275.
Raynaud, J.P., 1977, R 5020, a tag for the progestin receptor,
    in: "Progesterone Receptors in Normal and Neoplastic Tissues,"
    W.L. McGuire, J.P. Raynaud, and E.E. Baulieu, ed., Raven
    Press, New York, p. 9.
Raynaud, J.P., Martin, P.M., Bouton, M.M., and Ojasoo, T., 1978,
    11β-methoxy-17-ethynyl-1,3,5(10)-estratriene-3,17β-diol
    (Moxestrol), a tag for estrogen receptor binding sites in
    human tissues, Cancer Res., 38:3044.
Richardson, G.S., and MacLaughlin, E.T., 1978, "Hormonal Biology
    of Endometrial Cancer," Vol. 42, UICC Technical Reports
    Series, Geneva.
Robel, P., Levy, C., Wolff, J.P., Nicolas, J.C., and Baulieu, E.E.,
    1978, Réponse à un anti-oestrogène comme critère d'hormono-
    sensibilité du cancer de l'endomètre, C.R. Acad. Sci. Paris,
    287:1353.
Robertson, D.M., Mester, J., Beilby, J., Steele, S.J., and Kellie,
    A.E., 1971, The measurement of high-affinity oestradiol
    receptors in human uterine endometrium and myometrium, Acta
    Endocrinol., 68:534.
Sanborn, B.M., Kuo, H.S., and Held, B., 1978, Estrogen and proges-
    togen binding site concentrations in human endometrium and
    cervix throughout the menstrual cycle and in tissue from
    women taking oral contraceptives, J. Steroid Biochem., 9:951.
Schmidt-Gollwitzer, M., Genz, T., Schmidt-Gollwitzer, K., Pollow,
    B., and Pollow, K., 1978, Correlation between oestradiol and
    progesterone receptor levels, 17β-hydroxysteroid dehydrogen-
    ase activity and endometrial tissue levels of oestradiol,
    oestrone and progesterone in women, in: "Endometrial Cancer,"
    M.G. Brush, R.J.B. King, and R.W. Taylor, ed., Baillière-
    Tindall, London, p. 227.
Scublinski, A., Marin, C., and Gurpide, E., 1976, Localization of
    estradiol 17β dehydrogenase in human endometrium, J. Steroid
    Biochem., 7:745.
Seematter, R.J., Hoffman, P.G., Kuhn, R.W., Lockwood, L.C., and
    Siiteri, P.K., 1978, Comparison of [^{3}H] progesterone and
    [6,7-^{3}H]-17, 21-dimethyl-19-norpregna-4,9-diene-3,20-dione

for the measurement of progesterone receptors in human malignant tissue, Cancer Res., 38:2800.

Sutherland, R., Mester, J., and Baulieu, E.E., 1977, Tamoxifen is a potent 'pure' anti-oestrogen in the chick oviduct, Nature, 267:434.

Sweat, M.L., and Bryson, M.J., 1970, Comparative metabolism of progesterone in proliferative human endometrium and myometrium , Am. J. Obstet. Gynecol., 106:193.

Syrjälä, P., Kontula, K., Jänne, O., Kauppila, A., and Vihko, R., 1978, Steroid receptors in normal and neoplastic human uterine tissue, in: "Endometrial Cancer," M.G. Brush, R.J.B. King, and R.W. Taylor, ed., Baillière-Tindall, London, p. 242.

Trams, G., Engel, B., Lehmann, F., and Maass, H., 1973, Specific binding of oestradiol in human uterine tissue, Acta Endocrinol., 72:351.

Tseng, L., and Gurpide, E., 1972a, Nuclear concentration of estradiol in superfused slices of human endometrium, Am. J. Obstet. Gynecol., 114:995,

Tseng, L., and Gurpide, E., 1972b, Changes in the in vitro metabolism of estradiol by human endometrium during the menstrual cycle, Am. J. Obstet. Gynecol., 114:1002.

Tseng, L., and Gurpide, E., 1974, Estradiol and 20α-dihydroprogesterone dehydrogenase activities in human endometrium during the menstrual cycle, Endocrinology, 94:419.

Tseng, L., and Gurpide, E., 1975a, Induction of human endometrial estradiol dehydrogenase by progestins, Endocrinology, 97:825.

Tseng, L., and Gurpide, E., 1975b, Effects of progestins on estradiol receptor levels in human endometrium, J. Clin. Endocrinol. Metab., 41:402.

Tseng, L., Gusberg, S.B., and Gurpide, E., 1977, Estradiol receptor and 17β-dehydrogenase in normal and abnormal human endometrium, Ann. N.Y. Acad. Sci., 286:190.

Tsibris, J.C.M., Cazenave, C.R., Cantor, B., Notelovitz, M., Kalra, P.S., and Spellacy, W.N., 1978, Distribution of cytoplasmic estrogen and progesterone receptors in human endometrium, Am. J. Obstet. Gynecol., 132:449.

Verma, U., and Laumas, K.R., 1973, In vitro binding of progesterone to receptors in the human endometrium and the myometrium, Biochem. Biophys. Acta, 317:403.

Wiest, W.G., and Rao, B.R., 1971, in: "Advances in the Biosciences," G. Raspe, ed., Vol. VII, p. 251.

Young, P.C.M., and Cleary, R.E., 1974, Characterization and properties of progesterone-binding components in human endometrium, J. Clin. Endocrinol. Metab., 39:425.

EMBRYO-ENDOMETRIAL RELATIONSHIPS DURING IMPLANTATION

Viviane Casimiri and Alexandre Psychoyos

Laboratoire Physiologie de la Reproduction
E.R. 203, C.N.R.S.
Hôpital de Bicêtre (INSERM)
78, Av. Gal Leclerc
Bicêtre, 94270, France

INTRODUCTION

The process of ovum implantation in mammalian uterine endometrium differs among species. Although the morphological as
well as the physiological details vary, certain conditions are
common to all types of ovum implantation: the embryo must be at
the blastocyst stage, and the endometrium, after precise preparation by ovarian hormones, must have reached a stage of optimal
receptivity. Therefore, successful implantation requires precise
timing between the development and interactions of the ova and the
endometrium. Failure of synchronization of the different events
delays or prevents egg implantation.

Im most species, the embryo (morula stage) arrives in the
uterus 3 to 4 days after fertilization, and reaches the blastocyst stage some hours later: 4.5 days for humans (Hertig, et al.,
1954) and late on day 4 for rats (De Feo, 1967). The first
condition necessary for implantation is transport of the blastocyst to the appropriate uterine site for nidation. The blastocyst then orients its embryonic pole toward the antimesometrial
epithelium (near the median line in women; Hertig, 1968) and loses
its zona pellucida upon effective contact with the endometrium;
zona-free human blastocysts have been found in the uterus approximately 4.5 days after fertilization (Falck-Larsen, 1978).

The first attachment event of the blastocyst is apposition to
the luminal epithelium. In species with large blastocysts,
apposition might result essentially from growth of the embryo.

63

In species bearing small blastocysts, the uterine lumen narrows
and the walls compress the blastocyst against the endometrium
(Enders, 1976). Progressive adhesion of the trophoblast to the
uterus involves changes in the luminal epithelium surface and
establishment of intercellular contact. The microvilli of the
blastocyst cell membranes and the luminal epithelium flatten
and both surfaces become closely associated (Enders, 1976).

In many species, adherence is followed by invasion of the
endometrium by the trophoblastic cells. In species demonstrating
eccentric implantation, uterine epithelial cells are displaced,
degenerate and are phagocytized (day 6 in rats; Enders, 1976).
In the interstitial type, active trophoblast invasion and fusion
of trophoblastic and epithelial cells occurs (observed on day 11
in humans; Falck-Larsen, 1978). Trophoblast cells then migrate
into the stroma, where cells have undergone decidual differentiation, and surround maternal glands and vessels. At this stage
implantation is achieved and formation of the placenta begins.

Although the mechanism of egg implantation varies among
species, essential steps are common to all types of ova implantation. In humans, this mechanism is not well understood since
tissue samples from early stages are not easy to obtain. However,
the first stages of implantation are well known in small rodents,
and thus we will focus on the events associated with rat blastocyst implantation.

ONSET OF BLASTOCYST-ENDOMETRIUM INTERACTIONS

Early on day 5 of pregnancy, rat blastocysts become regularly
spaced in each of the uterine horns, probably as a result of
uterine muscular activity (O'Grady and Heald, 1969) and at approximately noon of the same day, they are found inside nidatory crypts
formed on the antimesometrial side of the uteri. The blastocysts
always adhere to the epithelium by trophoblastic cells that surround the embryonic mass (Psychoyos, 1973; Fig. 1).

In chis area, the well-developed blood capillaries become more
permeable and the stroma undergo a generalized edema (Psychoyos,
1973). As a result of progressing edema, the uterine lumen disappears and the narrowing endometrium compresses the blastocysts
between the epithelial microvilli (Psychoyos, 1967). Stromal
edema is a characteristic response of the uterus to possible
implantation. A similar edema occurs in women at approximately
the 20th day of the menstrual cycle (Noyes et al., 1950). It can
develop independently of the blastocyst presence in both humans
and small rodents.

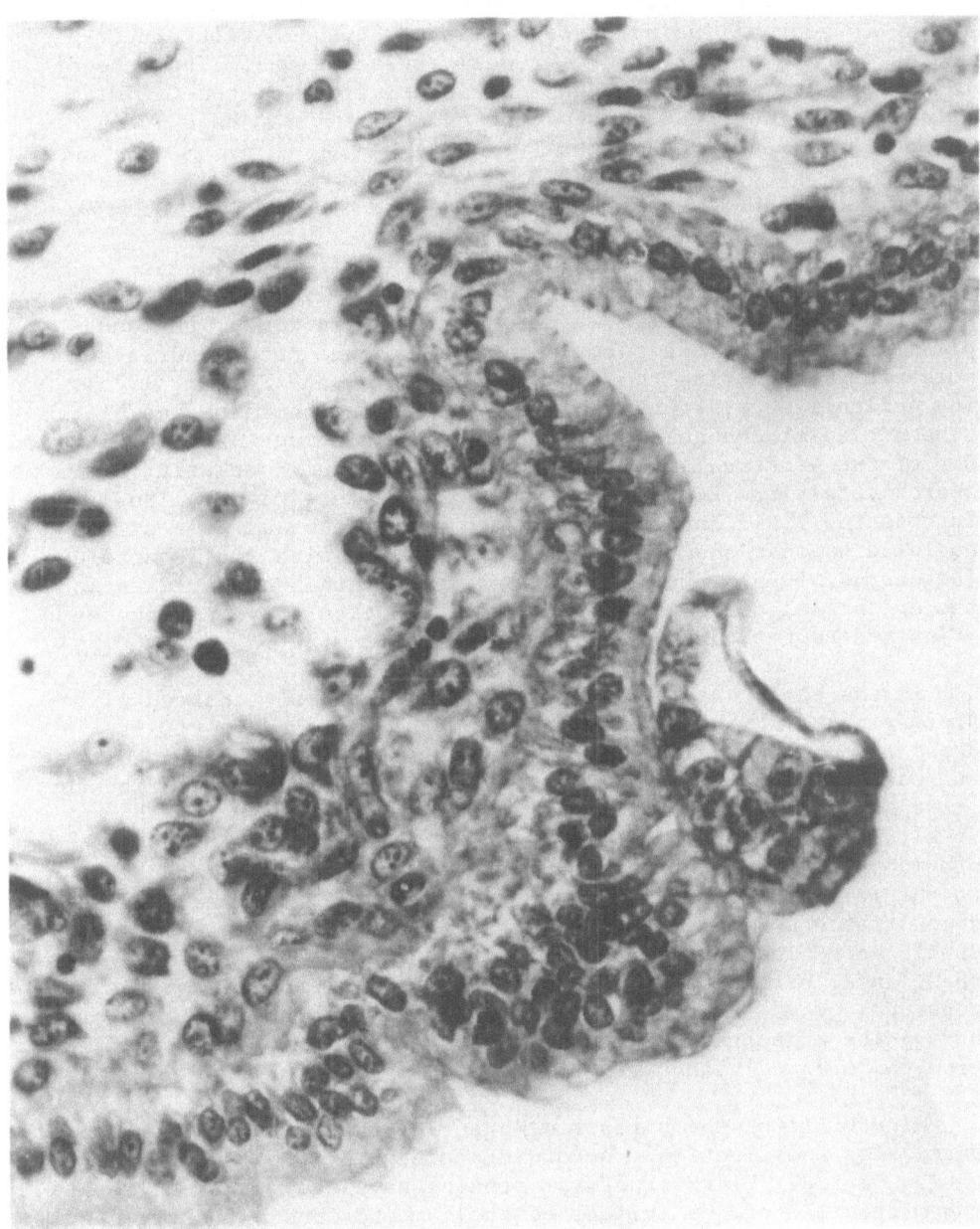

Fig. 1. Longitudinal section of a rat uterus on day 5 of pregnancy,
         showing apposition of the blastocyst to the antimesome-
         trial epithelium. Note stromal edema and the central
         position of the epithelial nuclei. x 400

In the rat, by early afternoon of the 5th day, the blastocyst loses its zona pellucida through blastocyst-endometrium interactions, which are not, as yet, well characterized. The role of the blastocyst may be a mechanical interaction (Bitton-Casimiri, et al., 1970), whereas that of the endometrium may be a lytic enzymatic action (Pinsker et al., 1974).

EPITHELIAL CHANGES

Numerous epithelial changes precede and accompany these events. Generally, the nuclei and the nucleoli increase in size and the inclusions increase in complexity. In women, the endoplasmic reticulum, the golgi apparatus and the mitochondria become more prominent during the luteal phase (Finn and Porter, 1975). The nucleoli show structures not observed in other species, i.e., "the nucleolar channel system". They are distinct from the 16th day of the menstrual cycle, reach their maximum complexity by approximately the 20th, and regress thereafter (Clyman, 1967). In rodents, lipid inclusions located near the basal pole of the nuclei disappear and the cytoplasm is filled with small vacuoles (Psychoyos, 1973). During this period, glycogen accumulates in the epithelial cells in other species, particularly in primates (Finn and Porter, 1975).

In the early stages of pregnancy, a negatively charged glycoprotein coat, which is considered part of the glycocalyx, is deposited on the microvilli (Hewitt et al., 1979). Recent data suggest that the glycocalyx, present on the epithelium during the first days of pregnancy, progressively disappears or its charge is altered at the time of implantation (Hewitt et al., 1979) providing favorable conditions for adherence of the trophoblast to the luminal epithelium (Enders, 1976). The glandular epithelium demonstrates concomitant intense secretory activity. An acidophilic secretion containing mucopolysaccharide and sulfate fills the glands, particularly in humans around the 20th day of the menstrual cycle (Finn and Porter, 1975). The fluid in the uterus during the progestational phase consists primarily of secretions from the uterine glands.

Microvilli of the surface membrane of epithelial cells undergo particular modifications, according to each step of the implantation period. Ultrastructural studies have shown that in rats the microvilli of the luminal epithelium are abundant and tentacle-like during early pregnancy (Nilsson, 1966), but by the afternoon of day 5, as the walls of the lumen begin to compress the blastocyst, the microvilli flatten and new structures appear. Most of the cells bear bulbous cytoplasmic projections with a depression at the center (pinopodes), which regress the next day (Psychoyos and Mandon, 1971; Psychoyos, 1973b). Similar structures have also been described in women by the 20th day of the menstrual cycle; these also regress rapidly thereafter (See chapter 2). The formation of epithelial pinopodes is, therefore, a transient phenomenon occurring only during the period of egg attachment (Fig. 2). It is thought that pinopodes may absorb fluid from the uterine lumen (Psychoyos, 1973b), thus facilitating closure of the lumen and apposition of the blastocyst to the luminal epithelium. They may also constitute the area of the first firm contact between embryonic and maternal cells. Blastocysts examined by scanning electron microscopy bear imprints in most of their cells indicating interdigitation of pinopodes and trophoblastic cells (Psychoyos, 1973b; Fig. 3).

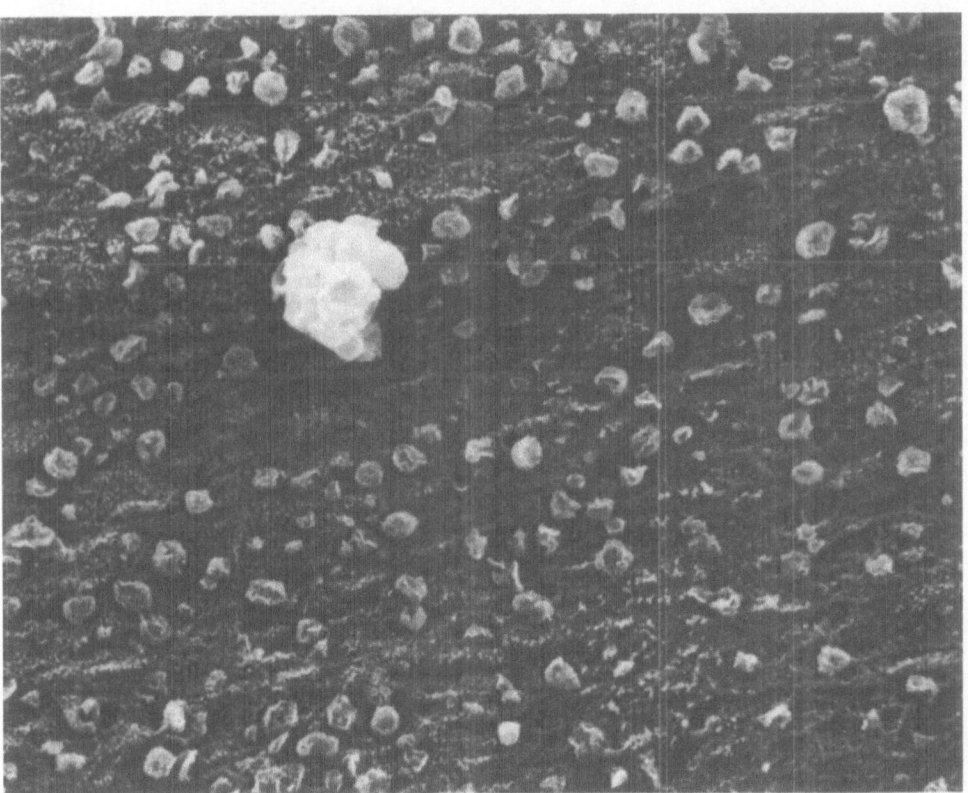

Fig. 2.   Scanning electron micrograph of rat luminal uterine surface during the early afternoon of the fifth day of pregnancy and part of an attached blastocyst. Note the abundance of protruding pinopodes on the surface of uterine epithelial cells. (x 1000)

Fig. 3. Same as Fig. 2, x 4000. Note imprints of epithelial pinopodes on the surface of blastocyst cells and organization of pinopodes protruding among the microvilli on the surface of epithelial cells.

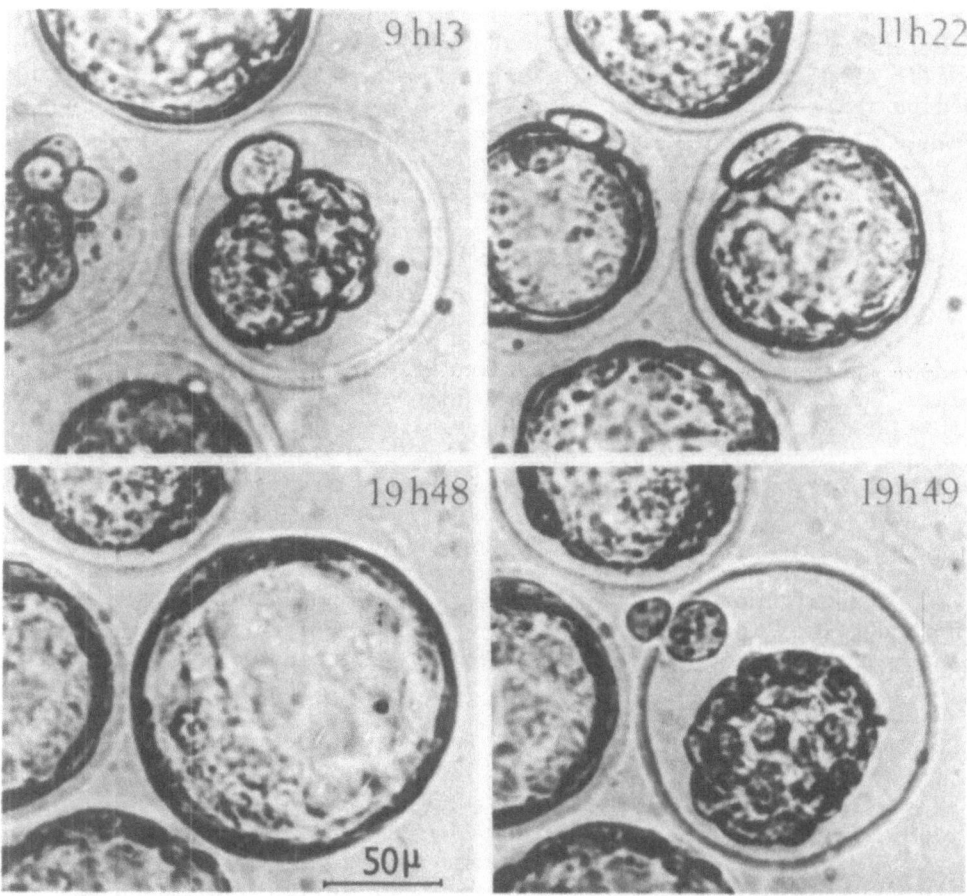

Fig. 4. Rat blastocysts developing in vitro as observed by time-
         lapse microcinematography. Note the rapid contraction
         of the blastocyst and emergence of the contents of the
         blastocoel as vesicles. The droplets are squeezed against
         the zona pellucida and emerge into the culture medium
         through local ruptures in the zona.

EMBRYONIC MESSAGE

Blastocysts observed in vitro by time-lapse microcinematography exhibit repetitive cycles of dilations and contractions (Bitton-Casimiri et al., 1970). These movements facilitate liberation from the zona pellucida and allow the emergence of blastocoelic droplets containing a granular material into the surrounding medium (Bitton-Casimiri et al., 1971; Fig. 4). If the blastocyst swells and shrinks in vivo as it does in vitro, the expelled blastocoelic fluid most probably is absorbed by adjacent pinopodes surrounding the blastocyst. Such material probably contains "messages", dispensed by the implanting embryo, that may be involved in immunorecognition, maintenance of the corpus luteum or stromal decidualization.

The nature of the embryonic message is not known; however, before the appearance of tight interdigitation between the trophoblast and epithelium surfaces, a dramatic increase in uterine capillary permeability occurs at the site of embryo-maternal contact (Psychoyos, 1960). This inflammatory-like reaction is the earliest response of the endometrium to the blastocyst or to other deciduogenic stimuli (for review see Psychoyos, 1973a). This response can be visualized in the pregnant rat from the evening of day 5 after intravenous injection of a macromolecular dye, i.e., Evan's blue or Pontamine blue. Blue spots due to locally increased capillary permeability indicate the location of the blastocysts (Psychoyos, 1973a). The first decidual cells appear several hours later.

DECIDUALIZATION

Decidual tissue formation is a specific characteristic of progestational endometrium. Stromal cells enlarge rapidly, lose their fibroblastic appearance, become polyploid, and accumulate glycogen, lipids and numerous enzymes (for review see Finn and Porter, 1975). Spontaneous decidualization occurs in human endometrium late in the luteal phase, regardless of blastocyst presence (Noyes et al., 1950). In contrast, in laboratory rodents this transformation appears naturally only during pregnancy and is induced by an unknown embryonic stimulus. However, a decidual reaction can also be induced artificially by a large variety of physical and chemical stimuli (Shelesnyak, 1957; De Feo, 1963) that mimic the effect of the blastocyst on the uterine epithelium.

Endometrial cells deprived of ovarian hormones do not decidualize. Progesterone treatment sensitizes the rat uterus to a deciduogenic stimulus of high intensity (i.e., trauma). However, estrogen is essential to reinforce endometrial sensitivity and to allow stimuli of low intensity (local chemical stimuli) to induce the reaction (De Feo, 1963). The hormonal requirements for induction of a decidual reaction by low intensity stimuli are identical to those necessary for an endometrial response toward the blastocyst (Finn and Porter, 1975).

HORMONAL CONTROL OF ENDOMETRIAL RECEPTIVITY

In rats under normal conditions, the period of maximal endometrial sensitivity is limited to several hours around noon on the fifth day of pseudopregnancy (De Feo, 1963). After this sensitive period, the endometrium no longer responds to deciduogenic stimuli (De Feo, 1963). Blastocysts transferred to uteri that have entered the nonsensitive state degenerate and are expelled (Psychoyos, 1973a). Thus, the brief receptive uterine phase that allows implantation and decidual reaction is followed by a state of "nonreceptivity" that is hostile to unimplanted embryos and refractory to decidual formation (Psychoyos, 1965).

The receptive and nonreceptive phases are induced by a precise progesterone—estrogen sequence, i.e., preparation of the endometrium by progesterone for at least 48 hours and intervention of a minute amount of estrogen (Psychoyos, 1965). The same basic progesterone-estrogen sequence that induces receptivity allows the uterus to enter a state of nonreceptivity.

Under a continuous progesterone regimen, the nonreceptive state is maintained indefinitely; however, if hormone treatment is stopped, the endometrium recovers its initial receptivity (Psychoyos, 1973b). The sequential changes in uterine receptivity can be controlled by exogenous hormones. For instance, in pregnant rats it is possible to inhibit implantation and decidual reaction by advancing the sensitive refractory stage with preovulatory progesterone treatment (Psychoyos, 1973a). Progesterone given systemically (Rudez et al., 1965) or released from capsules placed in the uterine cavity desynchronizes ovoendometrial evolution in several species, including women (Scommegna et al., 1970).

FACTORS INVOLVED IN ENDOMETRIAL REACTIVITY

Ovulatory hormone secretion, associated with embryonic stimuli, induces an inflammatory-like reaction characterized by increased endometrial capillary permeability at implantation sites. This reaction is the preliminary condition required for realization of egg implantation and decidual transformation. What makes the uterus sensitive and then insensitive to the blastocyst stimulus? The synthesis and release of vasoactive mediators may play an essential role in vascular permeability modulation. Histamine appears to be involved in endometrial vascular changes due to estrogen and/or to the blastocyst (Shelesnyak, 1957).

In rabbits, inhibition of histidine decarboxylase, the enzyme involved in de novo histamine synthesis, interrupts implantation (Dey et al., 1979). Furthermore, in this species the blastocyst exhibits a noticeable histidine decarboxylase activity (Dey et al., 1979). The synthesis of histamine, as a consequence of histidine decarboxylase activation near or in the endothelial capillaries via estrogen or other inducers, may be involved in increased endometrial capillary permeability. This system may also be responsible for human endometrial decidualization, which occurs normally every cycle. Nevertheless, other vasoactive substances, such as bradykinine, may also be implicated in vascular permeability and decidualization (Psychoyos, 1967).

The vascular response induced by vasoactive mediators is potentiated by prostaglandins (Vane, 1976). The uterine level of prostaglandins is maximal during the receptive period and if prostaglandin synthesis is inhibited by indomethacin, estrogen fails to induce implantation (Saksena et al., 1976). However, this effect can be overcome by the association of prostaglandin and histamine. Catecholamines and their receptors are certainly involved in the regulation of the uterine microcirculation. The activity of monoamine oxidase, the enzyme complex involved in deactivation of catecholamines, increases in human (Southgate et al., 1968) and rat endometrium during the luteal phase (Rath et al., 1979).

UTERINE FACTORS AFFECTING THE BLASTOCYST

Besides those described above, the luminal uterine fluid contains other factors that may be implicated in uterine receptivity or nonreceptivity to the blastocyst. Evidence supporting the existence of factors in the luminal uterine fluid affecting metabolism and survival of blastocysts has been demonstrated in several experiments. Blastocysts transferred to host uteri early in pregnancy may implant and develop, whereas blastocysts transferred after the period of uterine sensitivity (the nonreceptive state), do not implant and rapidly degenerate (Noyes et al., 1963; Psychoyos, 1973a). However, blastocysts can implant in ectopic sites (Kirby, 1967) or develop in vitro in relatively simple media (Bitton-Casimiri and Psychoyos, 1968).

The existence of a blastocyst metabolic inhibitor in the luminal fluid has also been studied in the nonreceptive uterus. The first studies examined the in vitro effect of uterine washings from rats under various hormonal conditions on incorporation of tritiated uridine by rat blastocysts. It was found that uterine washings collected during the nonreceptive period almost completely inhibited uptake and incorporation of [^{3}H]uridine by blastocysts incubated in these fluids (Fig. 5). Washings from ovariectomized

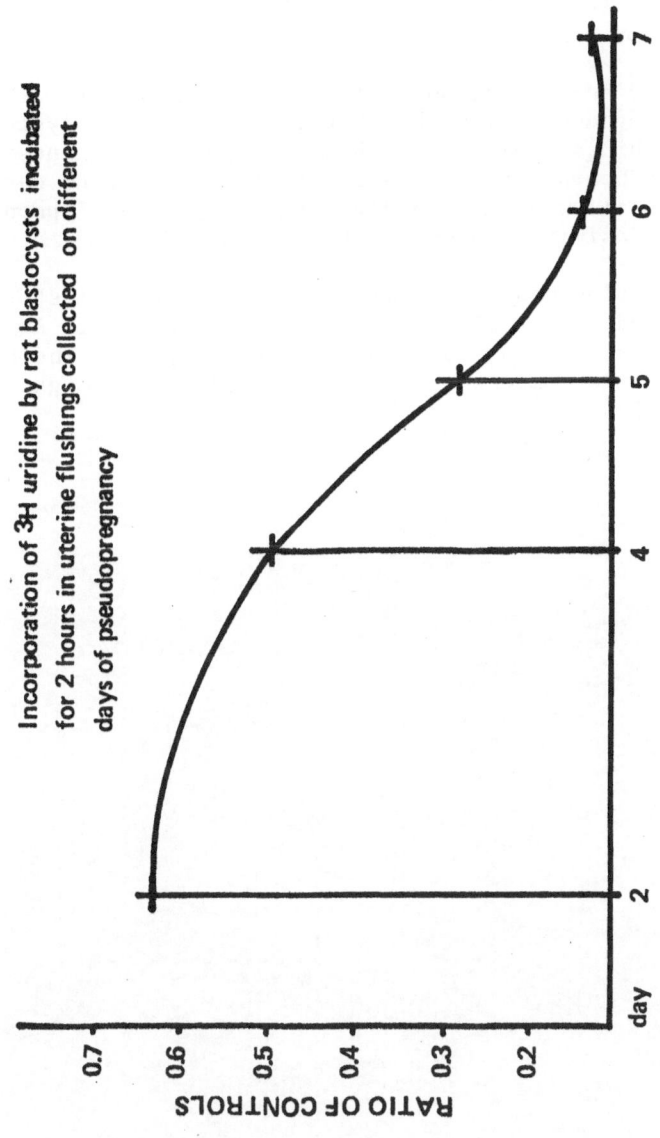

Fig. 5. Blastocysts were cultured 20 hours in normal medium; culture medium that had been flushed through the uteri of rats on the indicated days of pseudopregnancy was added, and 2 hours later blastocysts were pulse-labeled for 45 min with [^{3}H]uridine.

animals that were either treated or nontreated with progesterone, or those from the first 4 days of pseudopregnancy, only partially affected [^{3}H]uridine uptake and incorporation (Psychoyos et al., 1975; Psychoyos and Casimiri, 1980). These experiments indicate that nonreceptive uterine fluids contain components that inhibit embryonic metabolic activity <u>in vitro</u> (Psychoyos and Casimiri, 1980a,b).

Recent studies have demonstrated that washings from the 6th day of pseudopregnancy are highly blastotoxic. Blastocysts cultured in washings from the sixth day of pseudopregnancy degenerate within 24 hours, whereas washings from the second day of pseudopregnancy allow normal <u>in vitro</u> blastocyst development (Psychoyos and Casimiri, 1980a,b; Figs. 6 and 7).

Six-day uterine washings completely lose their blastotoxicity when dialyzed and permit blastocyst growth in culture. After dialysis, the toxic factor is found in the dialysate, indicating that the implicated component has a small molecular weight (Psychoyos and Casimiri, 1980a,b; Table 1).

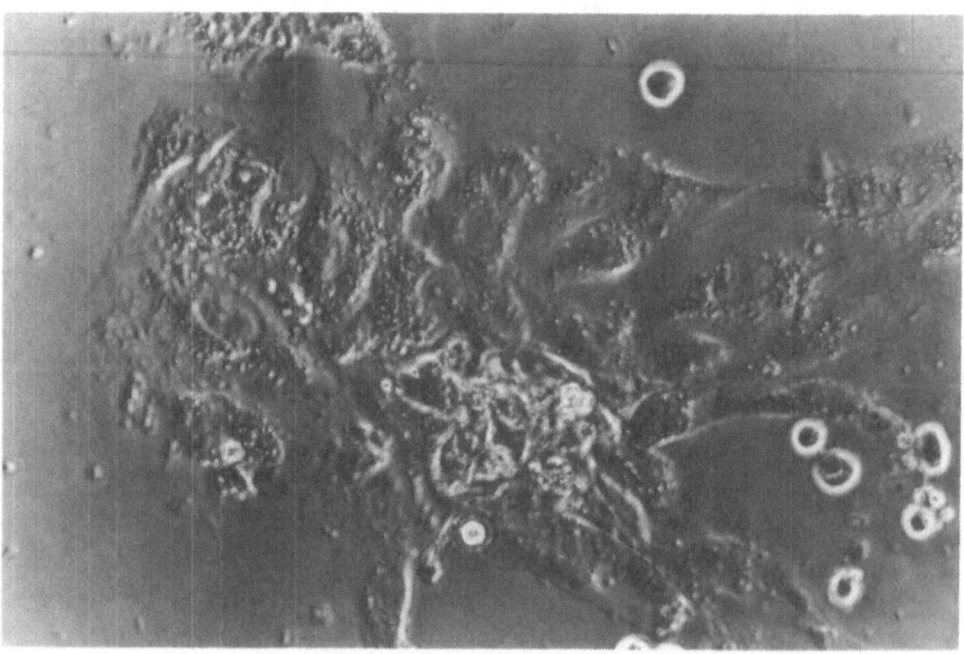

Fig. 6.   Rat blastocysts cultured 4 days in uterine washing
          collected on the second day of pseudopregnancy. This
          fluid allows blastocyst development and growth.   x 800.

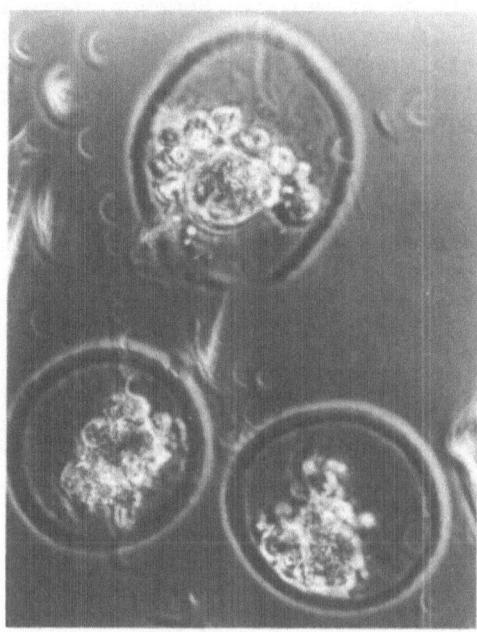

Fig. 7.   Rat blastocyst cultured in uterine washing collected on
           the sixth day of pseudopregnancy.   In this fluid, blasto-
           cysts degenerate within 24 hours.   x 400.

Further characterization was carried out by fractionation of
the day 6 dialysate on a Sephadex G–25/G–10.   By this technique,
six different fractions were obtained; the toxic component was
found to be restricted to Fraction IV (Table 1).   Blastocysts
cultured in this fraction or in nontreated washings degenerated
rapidly within 24 hours; none of the other fractions produced
this effect within this time period (Psychoyos and Casimiri,
1980a,b; Table 1).   The time of elution and the absorbance profile
of the blastotoxic component suggest that it may be a polypeptide
with a molecular weight of approximately 700 daltons (Psychoyos
and Casimiri, 1980a,b).   Until more information on the chemical
structure and biological properties of this blastotoxic component
isolated from nonreceptive rat uterus is obtained, we have termed
the factor blastocidin (Psychoyos and Casimiri, 1980a,b).

CONCLUSION

It is evident from our examination of the embryo-endometrial
relationships that several steps in the implantation process are
not yet clearly defined.   What enables apposition and attachment of
the blastocyst to the luminal epithelium?   What precise factor(s)
is involved in blastocyst interaction with the endometrium during
implantation?   What regulatory mechanisms control endometrial
reactivity?   These questions regarding the mysteries of egg

implantation remain to be answered. However, some elements of these phenomena have become clearer, specifically, the factors involved in endometrial vascular receptivity and substances, such as blastocidin, that may control intrauterine development and survival on the embryo.

Table 1. In Vitro Development of Rat Blastocysts Cultured for 24 Hours in Medium Containing Washings Collected on Day 2 (W2), or Day 6 (W6) of Pseudo-pregnancy, or Day 6 Washings Treated by Dialysis or by Fractionation on Sephadex G-25/G-10 Column[a]

Sample	Number of blastocysts	% Normal blastocysts	% Degenerated blastocysts
None	117	100	0
W2	38	100	0
W6	45	0	100
W6 dialyzed (A)	22	100	0
W6 dialyzate (B)	30	0	100
A + B	20	0	100
W6 fraction I	37	100	0
W6 fraction II	30	100	0
W6 fraction III	36	70	0
W6 fraction IV	34	0	100
W6 fraction V	30	90	0
W6 fraction VI	30	100	0

[a]Data from Psychoyos and Casimiri (1980b).

REFERENCES

Bitton-Casimiri, V., Brun, J.L., and Psychoyos, A., 1970, Comportement in vitro des blastocystes du 5e jour de la gestation chez la ratte. Etude microcinematographique, Compt. Rend. Acad. Sci., Paris, 270:2979.

Bitton-Casimiri, V., Brun, J.L., and Psychoyos, A., 1971, Active release of material from rat blastocyst developing in vitro, J. Reprod. Fertil., 271:461.

Bitton-Casimiri, V., and Psychoyos, A., 1968, Dévelopment du blastocyste du rat in vitro, Compt. Rend. Acad. Sci., Paris, 267:762.

Clyman, M.J., 1967, Fine structure of the nucleolus of the human endometrium at the time of nidation, Excerpta Med. Found. Intern. Congr. Ser., 133:400.

De Feo, V.J., 1963, Determination of the sensitive period for the induction of deciduomata in the rat by different inducing procedures, Endocrinology, 73:488.

De Feo, F.J., 1967, Decidualisation, in: "Cellular Biology of the Uterus," R.M. Wynn, ed., Appleton Century Crofts, N.Y.

Dey, S.K., Johnson, D.C., and Santos, J.G., 1979, Is histamine production by the blastocyst required for implantation in the rabbit? Biol. Reprod., 21:1169.

Enders, A.C., 1976, Anatomical aspects of implantation, J. Reprod. Fertil., [Suppl.], 25:1.

Falck-Larsen, J., 1978, Aspects ultrastructuraux de l'ovoimplantation humaine, in: "L'implantation de l'oeuf," F. du Mesnil du Buisson, A. Psychoyos, K. Thomas, ed., Masson, Paris.

Finn, C.A., and Porter, D.G., 1975, "The Uterus, Handbooks in Reproductive Biology," vol. 1, Elek Science, London.

Hertig, A.T., 1968, "Human Trophoblast," Charles C. Thomas, Springfield, Ill.

Hertig, A.T., Rock, J., Adams, E.C., and Mulligan, W.J., 1954, On the preimplantation stages of the human ovum, Contrib. Embryol., 35:199.

Hewitt, K., Beer, A.E., and Grinnell, F., 1979, Disappearance of anionic sites from the surface of the rat endometrial epithelium at the time of blastocyst implantation, Biol. Reprod., 21:691.

Kirby, D.R.S., 1967, Ectopic autografts of blastocysts in mice maintained in delayed implantation, J. Reprod. Fertil., 14:515.

Nilsson, O., 1966, Structural differentiation of luminal membrane in rat uterus during normal and experimental implantations, J. Anat., 125:152.

Noyes, R.W., Dickman, Z., Doyle, L., and Gates, A.H., 1963, Ovum transfers synchronous and asynchronous in the study of implantation, in: "Delayed Implantation," A.C. Enders, ed., University of Chicago Press, Chicago.

Noyes, R.W., Hertig, A.T., and Rock, J., 1950, Dating the endometrial biopsy, Fertil. Steril., 1:3.

O'Grady, J.E., and Heald, P.J., 1969, The position and spacing of implantation sites in the uterus of the rat during early pregnancy, J. Reprod. Fertil., 20:407.

Pinsker, M.C., Sacco, A.G., and Mintz, B., 1974, Implantation-associated proteinase in mouse uterine fluid, Devel. Biol., 38:285.

Psychoyos, A., 1960, Nouvelle contribution à l'étude de la nidation de l'oeuf chez la ratte, Compt. Rend. Acad. Sci., Paris, 251:3073.

Psychoyos, A., 1965, Control de la nidation chez les mammifères, Arch. Anat. Microsc. Morphol. Exp., 54:85.

Psychoyos, A., 1967, The hormonal interplay controlling egg implantation in the rat, in: "Advances in Reproductive Physiology," A. McLaren, ed., Logos Press, London.

Psychoyos, A., 1973a, Endocrine control of egg implantation, in: "Handbook of physiology; Endocrinology," R.O. Greep and E.B. Astwood, ed., Vol. II, Williams and Wilkins, Baltimore.

Psychoyos, A., 1973b, Hormonal control of ovoimplantation, Vitam. Horm., 31:201.

Psychoyos, A., Bitton-Casimiri, V., and Brun, J.L., 1975, Repression and activation of mammalian blastocysts, in: "Regulation of Growth and Differentiated Function in Eukaryotic Cells," G.P. Talwar, ed., Raven Press, New York.

Psychoyos, A., and Casimiri, V., 1980a, Factors involved in uterine receptivity and refractoriness, Prog. Reprod. Biol., 7:143.

Psychoyos, A., and Casimiri, V., 1980b, Uterine blastotoxic factors, in: "Cellular and Molecular Aspects of Implantation," S. Glasser and D. Bullock, ed., Plenum Publishing Corp., N.Y. (in press).

Psychoyos, A., and Mandon, P., 1971, Etude de la surface de l'épithelium utérin, au microscope électronique à balayage. Observations chez la ratte au 4e et au 5e jour de la gestation, Compt. Rend. Acad. Sci., Paris, 272:2723.

Rath, N.C., Olmedo, C., Casimiri, V., Parvez, S., Roche, D., Parvez, H., and Psychoyos, A., 1979, Monoamine metabolism during early pregnancy in the rat, in: "Research on Steroids," A. Klopper, L. Lerner, H.J. van der Molen and F. Sciarra, ed., vol. 8, Academic Press, New York.

Rudez, H.W., Martinez-Manautou, J., and Maqueo-Topete, M., 1965, The role of progestagens in the hormonal control of ferrility, Fertil. Steril., 16:158.

Saksena, S.K., Lau, I.F., and Chang, M.C., 1976, Relationship between oestrogen, prostaglandin F2α and histamine in delayed implantation in the mouse, Acta Endocrinol. (Kbh), 81:801.

Scommegna, A., Pandya, N.G., Lee, C.M., Christ, M., and Cohen, M. R., 1970, Intrauterine administration of progesterone by slow releasing device, Fertil. Steril., 21:201.

Shelesnyak, M.C., 1957, Aspects of reproduction. Some experimental studies on the mechanism of ovoimplantation in the rat. Recent Prog. Horm. Res., 13:269.

Southgate, J., Grant, E.C., Pollard, W., Prise-Davies, J., and Sandler, M., 1968, Cyclical variation in endometrial monoamine oxidase correlation of histochemical and quantitative biochemical assays, Biochem. Pharmacol., 17:721.

Vane, J.R., 1976, Prostaglandins as mediators of inflammation, in:
"Advances in Prostaglandin and Thromboxane Research," B.
Samuelson and R. Paoletti, ed., vol. II, Raven Press, New
York.

BIOCHEMICAL EVALUATION OF CORPUS LUTEUM FUNCTION

Robert Scholler, Khalil Nahoul, and Catherine Blacker

Fondation de Recherche en Hormonologie
26 Boulevard Brune
75014 Paris
    and
67 Boulevard Pasteur
94260 Fresnes (France)

INTRODUCTION

Based on the vast amount of data obtained from several animal species as well as in man (Csapo and Pulkkinen, 1978), corpus luteum function appears to be essential for implantation of the ovum and development and maintenance of early pregnancy up to the seventh week (Csapo et al., 1972). Moreover, abortion following removal of the corpus luteum can be prevented by progesterone (P) administration, whereas 17β-estradiol is inefficient (Csapo et al., 1973a,b).

The indispensability of P in early pregnancy maintenance was clearly demonstrated as early as the 1940's upon recognition that levels of pregnanediol glucuronide, a major P metabolite found in urine, decreased in threatened abortions (Venning et al., 1937). P is also necessary for ovum implantation, as recently shown by Csapo and Resch (1979) who observed that isoxazol, an antiprogesterone synthetic steroid, prevents implantation in rats, whereas concomitant P administration counteracts this effect. P thus appears to be the most important hormone of the corpus luteum, its effects mimicking those of the corpus luteum as a whole.

Therefore, luteal phase defects may be defined as impaired P production by the corpus luteum (Jones, 1976) so that it is necessary to evaluate this production to diagnose any luteal disorder. This assessment can be made by P production rate determin-

ation; however, this method is much too cumbersome and time con-
suming in clinical practice. Plasma P concentrations have been
shown to be a good index of the P production rate throughout the
luteal phase (Baird, 1976) and can now be easily measured by radioimmunoassay. However, daily plasma P levels can only be a reliable
diagnostic index when compared to the normal range established in
a group of strictly selected normal subjects.

    This paper first reviews the hormonal features of the normal
cycle, with particular emphasis on the mode of calculating the
normal range of plasma P levels throughout the luteal phase.
Hormonal findings in luteal phase defects are examined later.

HORMONAL PATTERNS THROUGHOUT NORMAL MENSTRUAL CYCLE

    Radioimmunoassays have enabled accurate establishment of
hormone patterns throughout the menstrual cycle. In 1978, Scholler
et al. published a thorough review on this subject; therefore,
hormone variations characteristic of the menstrual cycle will
be only outlined here (Fig. 1 A–F).

Gonadotropins

    The lowest levels of follicle stimulating hormone (FSH) and
luteinizing hormone (LH) occur approximately 10 days after the
midcycle peak. Thereafter, the two gonadotropin profiles diverge.
Whereas FSH exhibits a secretory wave ending at a relatively low
point a few days before the preovulatory peak, LH rises gradually
and steadily to the characteristic preovulatory surge, so that at
the end of the follicular phase, the decreasing FSH curve crosses
the rising LH curve. Thus, contrary to what might be expected,
the gonadotropin cycle does not start with the onset of the menses,
but begins and ends about 9 to 10 days after the midcycle peaks.

Estrogens

    Closely related to gonadotropin patterns, plasma estrogens
display two maxima, measurable in urine. FSH and 17β–estradiol
(E_2) curves are mirror images of each other throughout the follicular and luteal phases, indicating that FSH secretion may be
regulated by negative E_2 feedback. Conversely, LH and E_2 profiles
are parallel after the third day of the cycle, which may be the
result of LH action on ovarian steroidogenesis. E_2 secretion
increases as LH levels rise to a maximum at midcycle, and triggers,
by positive feedback, the release of LH/FSH–RF, which is responsible for the LH and FSH surges that generally occur 24 hours after
the E_2 peak. The Estrone (E_1) pattern is very similar to that of
E_2, but to a lesser degree and at lower levels. This is not
surprising since E_1 originates from various sources, whereas E_2

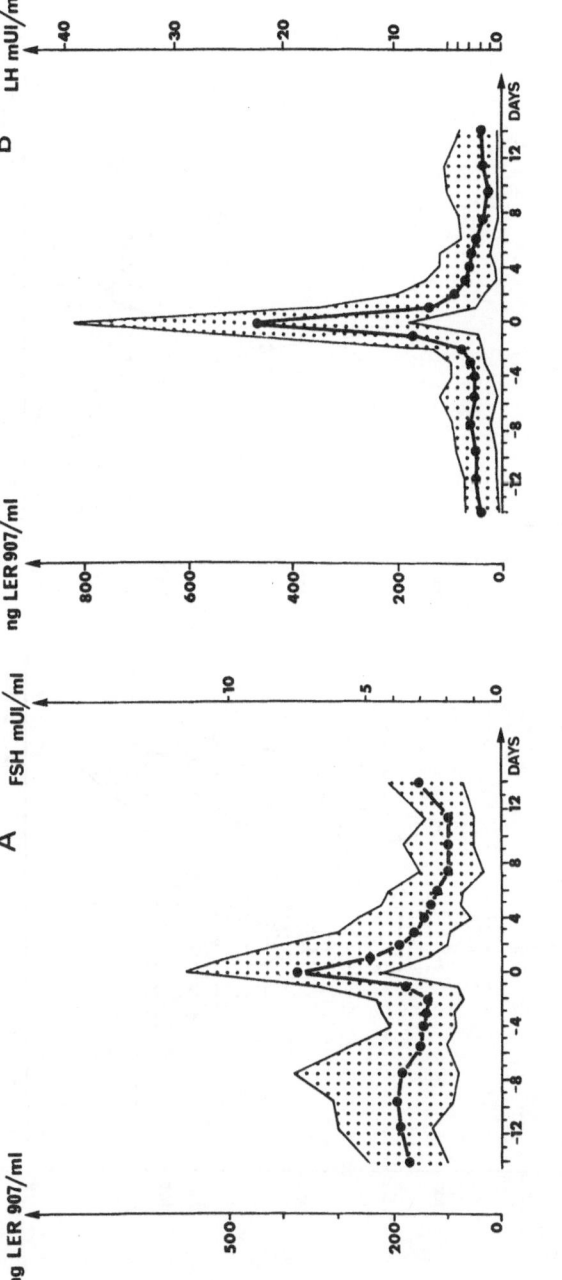

Fig. 1 A-B. Plasma hormone patterns throughout the menstrual cycle (Day 0 = day of the LH peak); A, FSH and B, LH.

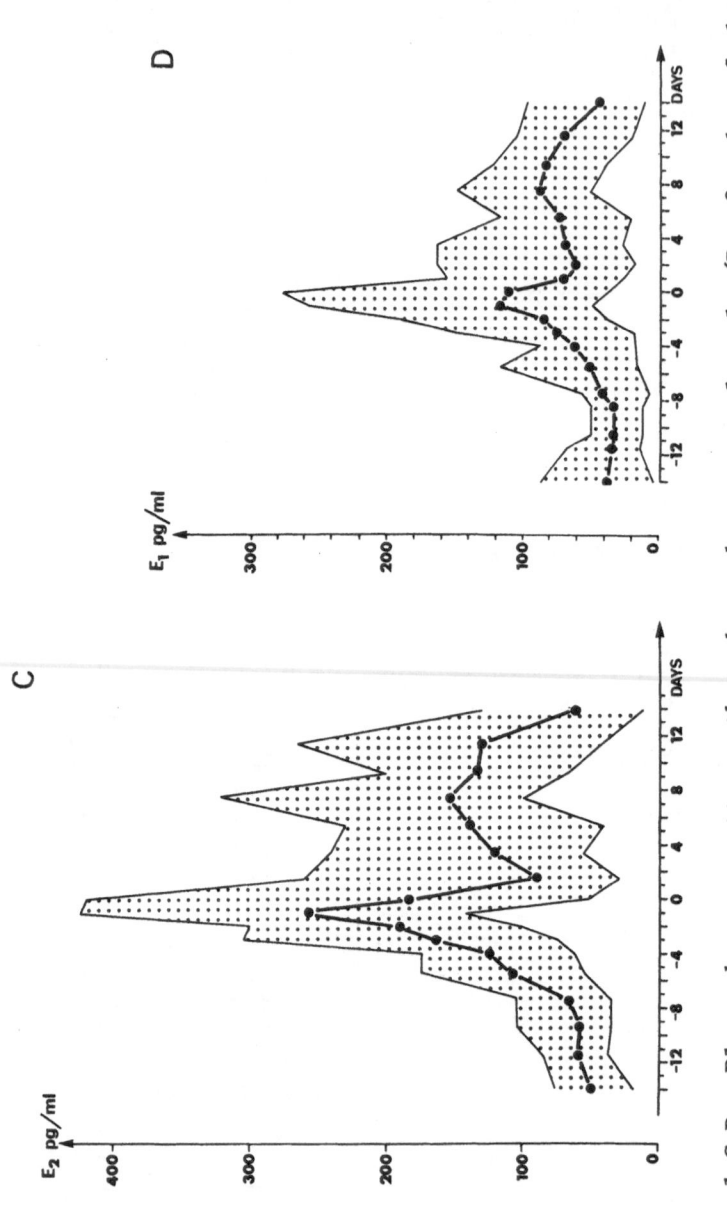

Fig. 1 C–D.   Plasma hormone patterns throughout the menstrual cycle (Day 0 = day of the LH peak); C, E_2 and D, E_1.

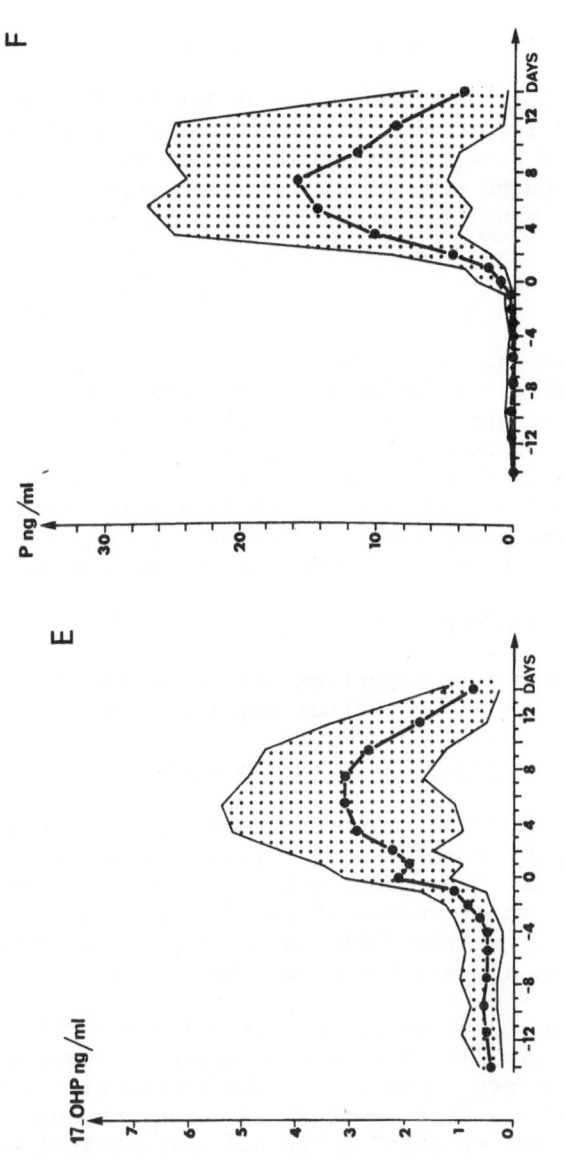

Fig. 1 E-F. Plasma hormone patterns throughout the menstrual cycle (Day 0 = day of the LH peak); E, 17-OHP and F, P.

is secreted primarily by the ovaries. Estriol (E_3) levels are very low throughout the menstrual cycle, although a significant difference has been demonstrated between the follicular and the luteal phases.

17-Hydroxyprogesterone (17-OHP)

During the follicular phase, 17-OHP levels are low, similar to those observed in prepubertal females, and persist until 4 days before the LH peak. Thereafter, 17-OHP rises to a maximum on the day of LH surge and, after a slight decrease, rises again during the luteal phase to levels higher than its preovulatory maximum.

Progesterone

Very low P levels, generally at the detection limit of current techniques, are observed during the follicular phase. However, a significant rise occurs on the day of the LH peak, indicative of the beginning of luteinization that precedes rupture of the mature follicle. P then achieves a luteal maximum 80 times higher, on the average, than its level during the follicular phase. The slope of the rising curve differs, however, from women to women, making it difficult to predict when the luteal peak will occur.

DETERMINATION OF NORMAL P RANGE

Because P is the most important hormone produced by the corpus luteum, its pattern will be further detailed.

Variability of P Pattern during Menstrual Cycle

The progesterone level patterns described throughout the menstrual cycle represent the mean obtained in a group of 34 women. In fact, wide individual variations were demonstrated. However, three different curves (Figs. 2-4) were visualized, based on data obtained from daily blood samples of 15 women and from Abraham et al. (1974); therefore, these data were pooled.

In the first group of 45 women, approximately 29% demonstrated an asymmetrical P profile, with a peak closer to ovulation than to the onset of the next menses. In the second group, 16 women or 35.5% displayed a rather symmetrical curve, whereas in the third group, 16 women or 35.5% again had asymmetrical curves, with a peak occurring rather late, closer to the onset of menses.

Statistical Studies

The above data were obtained in longitudinal studies, whereas in routine clinical investigations a limited number of blood

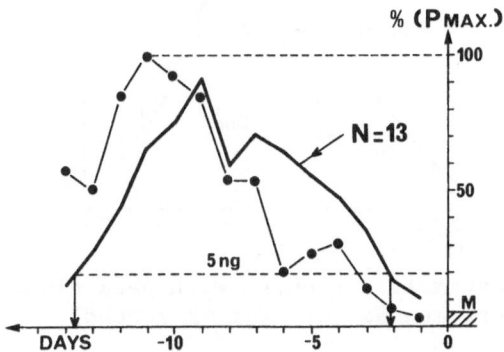

Fig. 2.   Type of P curve with peak closer to ovulation. The
curve without points corresponds to the average, whereas
the other represents an individual case. To permit
comparison of curves, the highest point of individual
curve has been defined as equal to 100.

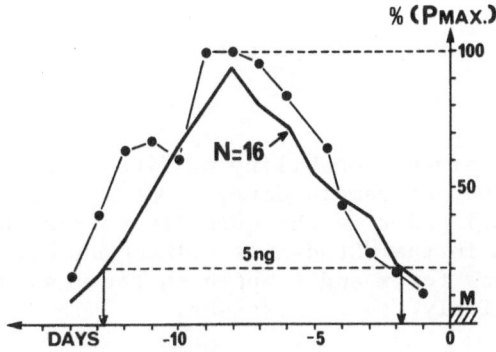

Fig. 3.   Type of symmetric P curve (cf. Fig. 2 legend).

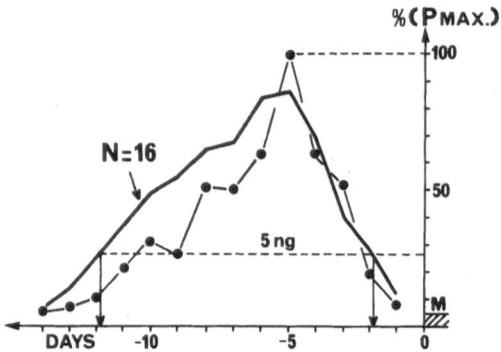

Fig. 4.   Type of asymmetric P curve with peak closer to the onset
of the next menses (cf. Fig. 2 legend).

samples is available. Thus, it is necessary to define the normal
range for each day of the luteal phase. Normal range limits are
diversely defined in the literature, generally because of postu-
lating a Gaussian distribution of the data and misusing the
parameters obtained. Indeed, when biological data are distributed
according to the Laplace-Gauss law, statistical analysis is greatly
facilitated. The mean and the standard deviation of the population
from which the observed sample is drawn can be easily estimated.
From the sample parameters:

$$\bar{x} = \Sigma xi/n \quad \text{and} \quad SD = \sqrt{\frac{\Sigma \, (xi - \bar{x})^2}{n-1}}$$

a confidence limit can be calculated:

$$\bar{x} \pm ts/ \sqrt{n}$$

where   there is a given probability of finding the population
mean; n is the number of sample data, $s/\sqrt{n}$ is the standard error
of the mean (S.E.M.) and t is the quantity corresponding to n - 1
degrees of freedom in the "Student's t distribution". These
unbiased estimations for $\bar{x}$ and s approach the mean and standard
deviation, respectively, as n increases.

     Sometimes, even when the hypothesis of a normal distribution
is correct, confusion between the statistics of the population
and those of the sample results in an incorrect definition of the
normal range limits. $\bar{x}$, S.E.M. and t can only be used to determine
the confidence limits of the mean, as outlined above, e.g., the

mean has a 95% chance of appearing in the range $\bar{x} \pm 1.96\ s/\sqrt{n}$, if n is large enough to compare sample means. $\bar{x}$ and SD are available to calculate the normal range limits of the population from which the sample has been drawn, if a k greater than t is substituted for t (Reed et al., 1971; Scholler, et al., 1973a,b, 1974).

In fact, biological data distributions are seldom Gaussian and are more frequently asymmetrical (Reed et al., 1971; Scholler et al., 1973a,b, 1974; Kletsky et al., 1975). Thus, using the Laplace-Gauss law to calculate normal range limits would lead to erroneous values, since the lower limit would sometimes be negative (Fig. 5). To prevent this drawback, most authors represent the mean curve crossed by vertical bars equivalent to $\pm$ 1 estimated SD. Some, however, given the non-Gaussian nature of the data distribution, assume that the distribution is log-Gaussian, since the data are asymmetrically distributed and fulfill some criteria of such a law. This was recently applied to plasma steroids by Kletsky et al. (1975). Such as assumption cannot be justified by the various tests available; thus, the hypothesis of such a distribution can only be accepted at a certain probability threshold. It must be emphasized that the normal range limits are closer to the reality; the lower limit is never negative (Fig. 5).

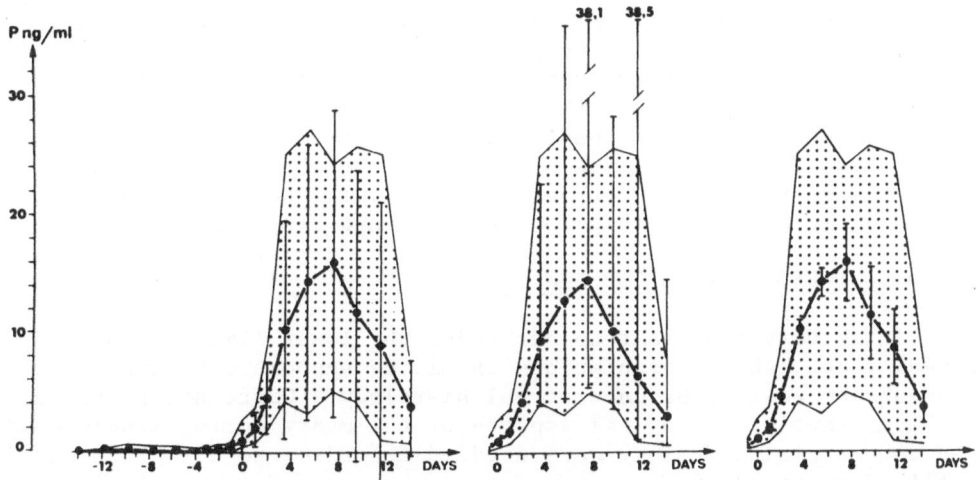

Fig. 5.   Different mode of calculation of P normal range throughout the luteal phase: a) mean $\pm 2 s_n$ Gaussian distribution; b) mean $\pm 2 s_{Ln}$ log-Gaussian distribution; c) mean $\pm$ SEM. (see Table 1).

In view of all these difficulties, it is preferable to resort to nonparametrical methods not requiring an a priori knowledge of the distribution type. In our approach, the lower limit is usually at the 5th percentile level (P5) and the upper limit is at the 95th percentile (P95), so that the normal range would include only 90% of the population. Obviously it would be more advisable to take the 2.5th and the 97.5th percentiles, but in this case, the sample size would not generally be large enough to calculate the confidence limits of these percentiles. As far as P is concerned, such a procedure is necessary for the 5th percentile, since it is most important to detect any defect in P production during the luteal phase. At the 90% probability threshold, the confidence interval of the P5 is delimited by two values, $P5_U$ and $P5_L$. Accordingly, there is a 90% chance that the 5th percentile of the population will fall within the interval thus defined. If the P level is above $P5_U$, there is a good chance that it is normal. If it is between $P5_U$ and $P5_L$, some uncertainty exists, and below $P5_L$, it is very likely to be insufficient.

The drawback of such methods is the necessity for a large sampling, generally over 120. Therefore, if the sample size is smaller, it is advisable to give the mean and the extreme values of the sample until completion of the study.

The normal range limits of plasma P levels obtained between days M-11 and M-4 (M = first day of the next menses) in our normal subjects, together with those calculated by Abraham et al. (1974), are reported in Table 1. As observed, there is good agreement between the data of these two groups, despite their different origins. According to our results, a P level above 6.8 ng/ml suggests a normal corpus luteum, whereas a level below 5 ng/ml strongly suggests luteal insufficiency. This lower limit (5 ng/ml), represented as a dashed line parallel to the abscissa in Figures 2-4, crosses the three P curves, respectively, on days M-14 and M-2 (Fig. 2), M-13 and M-2 (Fig. 3) and M-12 and M-2 (Fig. 4).

Anatomical-Biochemical Correlations

It is very important to determine the lower limit of the normal range; these calculations should be supported by anatomical findings as well as by endometrial histology. Correlations between P plasma levels, anatomical aspects of the ovary during laparoscopy and endometrial histology were studied according to the protocol outlined in Figure 6.

A good correlation was observed between plasma P levels and the presence or absence of a corpus luteum. Moreover, when blood samples were taken between days M-11 and M-4, a normal corpus luteum corresponded to a P level above 5 ng/ml (Scholler et al., 1978). Conversely, in some cases there was disagreement between P

Table 1. Calculation of Mean and Limits of Normal
Range for Plasma P (ng/ml)

	Abraham et al. (1974)	Scholler et al. (1978)
Number of women	30	189
Days of cycle[a]	M-11 to	M-4
Measurements	240	234
Mean ($\bar{x}$)	15.9	14.4
Normal Range[b]		
Gaussian Distribution $\pm ks_n$	0.8 - 31.0	3.0 - 25.8
Log-Gaussian Distribution $\pm ks_Ln$	4.8 - 41.7	5.6 - 31.8
5th Percentile (P5)	5.2	5.7
Confidence Level P5 (P=90%)	4.2 - 6.2	5.0 - 6.8

[a]M = Day of onset of menses following assay.
[b]For statistical definitions, cf. text; k≈2.1
s_n = Estimate of the standard deviation assuming a Gaussian distribution, s_{Ln} = Estimate of the standard deviation assuming a log-Gaussian distribution.

levels and the histological aspect of the endometrium, as already observed by others (Cooke et al.,1972; Radwanska and Swyer, 1974; Rosenfeld and Garcia, 1976; Shepard and Senturia, 1977; Annos et al., 1980). Rosenfeld and Garcia (1976) have suggested that the association of proliferative endometrium with a plasma P level greater than 3 ng/ml might be due either to inadequate endometrial response secondary to poor estrogen sampling; however, no explanation concerning the occurrence of secretory endometrium associated with low plasma P level (less than 2 ng/ml) was provided. Because blood collection and endometrial biopsy are generally performed concomitantly, some of the discrepancies might be more simply explained by the fact that the P level provides only instantaneous information on corpus luteum function, whereas the endometrium reflects the action of all hormones secreted during the previous part of the cycle. Moreover, a wide range of P levels corresponds to each day of histological dating.

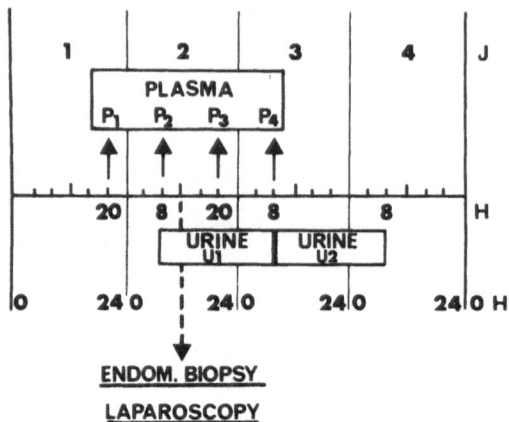

Fig. 6. Protocol for studying anatomical-biochemical correlations.

Practical Considerations

        Since the major source of P is the "active" ovary during
the luteal phase (Baird, 1976), corpus luteum function ideally
should be evaluated by determination of the P production rate,
but this is not feasible. As the P metabolic clearance rate does
not undergo significant variations during the luteal phase, the
plasma P determination can provide an accurate measurement of P
production. The occurrence of inter- and intraindividual variations makes it necessary to evaluate daily plasma P levels;
however, such an approach is cumbersome, expensive and uncomfort-
able for the patient, and only limited numbers of blood collections
are available in clinical routine. Once the normal range has been
clearly defined, it can be assumed that a single blood sample
collected in the middle of the luteal phase is a reliable index of
P production by the corpus luteum. In fact, a single plasma P level
can only evidence the existence of corpus luteum if it is above
5 ng/ml, the lower limit of the normal range calculated by nonparametrical tests and also reported by Ross et al. (1970) and
Abraham (1974). Even if the P level is higher than 10 ng/ml, the
lower limit of the luteal plateau, according to Jahansson (1969)
and more recently to Hensleigh and Fainstat (1979), it cannot
provide an accurate assessment of the adequacy of corpus luteum
function since the P pattern varies among patients and the normal

range is very wide. In fact, the P level may be greater than the lower limit of the normal range even though the menstrual cycle is abnormal (Abraham et al., 1974).

To circumvent these problems, Abraham et al. (1974) suggested collecting three blood samples between days M-11 and M-4. In normal cycles, the sum of the three P levels should be equal to or greater than 15 ng/ml. Indeed, in abnormal cycles the sum was less than this value, although individual values were higher than 3 ng/ml, the lower limit associated with a biphasic basal body temperature (BBT) chart and secretory endometrium according to Israel et al. (1972). When we tested this criterion in normal cycles, it was observed that the sum of the three P levels obtained between days M-11 and M-4 was equal to or greater than 15 ng/ml. This was also demonstrated even if the three lowest levels for each cycle were taken into account (Table 2; Scholler et al., 1977).

Table 2.   Three Lowest P Levels (ng/ml): Sum Between Days M-11 and M-4 (M=First Day of Next Menses)

	Abraham et al. (1974)	Scholler et al. (1977)
Mean	32.3	34.7
Lower Limit	15.1	14.8

Though this approach has proven useful in general, it should be noted that it might be more interesting to measure the area below these three points. Yet it remains to be demonstrated that this area would be equivalent to the more sophisticated "luteal index" of Del Pozo et al. (1979). This index is determined by integration of the area below the curve of the plasma P levels obtained every second day throughout the postovulatory period.

LUTEAL PHASE DEFECT

On the basis of BBT charts, pregnanediol excretion and endometrial biopsy, Jones (1949) first described luteal phase defects in some infertile patients with ovulatory cycles. Since then, many studies have been devoted to this pathological condition (see reviews Jones, 1976; Andrews, 1979) and according to these data, luteal defects can be defined as an impairment of P production by the corpus luteum.

Clinically, different degrees of luteal defect may be observed, the most severe form being the aluteal cycle. The short luteal phase and the inadequate luteal phase (also called luteal phase defect or corpus luteum insufficiency) need not be distinguished as two clinical entities, as they represent the severe and the moderate form of the same disorder, respectively. Diagnosis of such disorders is most important since they are considered to prevent successful gestation by interfering with conception and implantation or by inducing early abortions (Jones, 1976; Hensleigh and Fainstat, 1979; Crépin et al., 1980; Scholler et al., 1980) or occult abortions that generally are undiagnosed because they occur near the expected date of the next menses (Soules et al., 1977; Seppälä et al., 1978).

The incidence of luteal defects in infertile women has been reported to vary from 3.5% (Jones, 1975; Murthy et al., 1970) to 8.1% (Rosenberg et al., 1980) and even to 20% (Gautray et al., 1978). In women with histories of recurrent abortions, a corpus luteum defect is observed more frequently (Jones, 1976; Horta et al., 1977; Yip and Sung, 1977; Hensleigh and Fainstat, 1979). However, the existence of luteal deficiency in 35% of habitual abortion patients, as reported by Jones (1976) must be accepted with caution since this high percentage was observed in a group more likely to have endocrine than other abnormalities.

Moreover, the resulting imbalance between progesterone and estrogens in luteal phase defects is claimed to be favorable for the development of breast (Sherman and Korenman, 1974c) and endometrial (Gambrell, 1979) carcinomas; however, no abnormalities in P production have yet been reported. This hypothesis still needs confirmation. The existence of such an imbalance in benign breast diseases continues to be debated (Mauvais-Jarvis et al., 1976; Sitruk-Ware et al., 1979; Geller et al., 1979).

Diagnostic Procedures

BBT recording. Based on the thermogenic properties of P, BBT measurement is the most commonly used method for the demonstration of ovulation (Geller, 1961). Luteal phase defects may be suspected on the basis of a short thermal shift on the BBT chart or a slow rise in temperature during the luteal phase. Diagnosis cannot, however, be established by this procedure since it is well known that many patients are poor temperature takers and that the degree of thermal response to P varies among individuals. Indeed, a monophasic BBT graph may be observed in women with documented ovulatory cycles. This was found in 20% of the cycles studied by Moghissi (1976) and in 12% of those studied by Johansson et al. (1972); these two percentages are, however, higher than those recently reported by Hilgers and Bailey (1980) who found a monophasic

BBT chart in only 2.7% of ovulatory cycles. In addition, these authors observed that in 6.8% of the cases, this procedure gave incorrect information on the ovulatory status of the cycle.

Assuming that temperature was recorded accurately throughout the menstrual cycle, the occurrence of the lowest point before the luteal shift varies with respect to the LH peak. In an earlier study (Scholler et al., 1978), we have shown that in 7 of 27 cases (22.2%) the day of the LH peak coincided with the lowest point of the BBT chart before luteal shift, whereas in 37% of the cases, the LH peak occurred the day before. In the remainder of the cycles, the lowest point of BBT took place either 2 days or up to 4 days after the midcycle LH peak. These findings, in agreement with Dhont et al. (1974), demonstrate that BBT recording may be mis-leading in assessing luteal phase duration. Moreover, a P level as low as 2.5 ng/ml is enough to induce an increase in the BBT (Ross et al., 1970), but, as shown, is insufficient, so that a biphasic BBT chart does not necessarily rule out an abnormal corpus luteum and low plasma P concentration.

Endometrial biopsy. Because endometrium is the target tissue of the steroids secreted by the corpus luteum, endometrial biopsy during the late luteal phase reflects both the amount and the duration of corpus luteum steroid production and, therefore, is the only diagnostic tool for luteal disorders recommended by Jones, (1976). Indeed, there is a good correlation between the dating of the endometrial biopsy (Noyes et al., 1950) and the date of ovu-lation, as determined by the LH peak (Tredway et al., 1973).

Diagnosis of luteal defects (Jones, 1976) can be established if the histology is out of phase by two or more days with the date of ovulation and with the onset of the next menstrual period. This criterion has been questioned by Tredway et al. (1973), since in 5 of 11 patients with normal menstrual cycles, histologic pat-terns were two days behind the expected date of ovulation, as determined from the LH peak. Recently, Rosenfeld and Garcia (1976) showed that endometrial biopsy should be coupled with plasma P determination to diagnose luteal phase deficiencies. Endometrial biopsy may, however, be useful to demonstrate a possible, although rare, progesterone receptor defect (Keller et al., 1979).

Hormonal Features of Luteal Phase Defects

As mentioned, plasma P determination based on a single blood sample, even though well timed during the midluteal phase, is not adequate to evaluate P production by the corpus luteum. A P level higher than the lower limit of the normal range, as calculated above, only suggests the existence of a corpus luteum. Even a P level of 15 or 20 ng/ml cannot exclude a luteal defect since this high level might not be sustained and the entire P production might be reduced (Del Pozo et al., 1979).

The criterion used by Abraham et al. (1974) implies that the sum of three P levels between days M-11 and M-4 must be equal to or greater than 15 ng/ml. In our opinion, each of these three levels should be taken into account so that the shape of the P curve can be determined.

It should be noted that sampling days can be determined according to the expected menses only in women with regular cycles. Otherwise, the BBT chart should be taken into consideration and blood withdrawn on the 3rd, 6th and 9th days of the thermal plateau; this dating has generally proved useful in our hands. P levels can be interpreted only in relation to the onset of the next menstrual period.

In the two particular cycles illustrated in Figure 7, the P level in the midluteal phase is within the normal range; even in 7a it is above 10 ng/ml and the sum of the three levels between

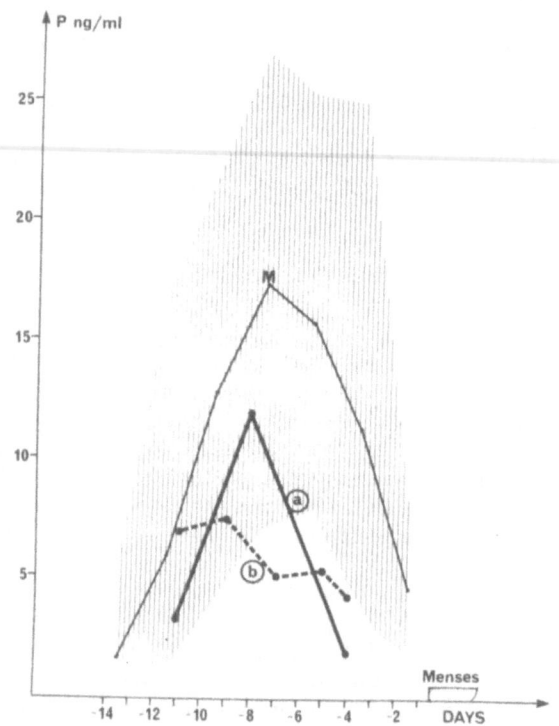

Fig. 7. P pattern in two individual cycles with luteal defect.

days M-11 and M-4 is greater than 15 ng/ml; however, the area
(49.8 and 41.9) is below the lower limit observed in our normal
group on the same sampling days (65.6). Once a luteal phase defect
is diagnosed by P determination, other hormones should be assessed
to establish the degree of severity and/or the etiology of the
disorder.

        In the short luteal phase, the menstrual cycle length may be
normal (Sherman and Korenman, 1974a) or decreased (Strott et al.,
1970). The minimum duration of normal luteal span has been defined
differently, i.e., 12 days by Del Pozo et al. (1979) and Soules
et al. (1977), 11 days by Ross et al. (1970) and Lehtovirta et al.
(1979), 10 days by Sherman and Korenman (1974a), Murthy et al.
(1970) and Quagliarello and Weiss (1979) and 8 days by Strott
et al. (1970) and Ross (1976). Such variations may be partly due
to the criteria selected for the determination of the first day
of the luteal phase. In this particular form of luteal phase
defect, many hormone abnormalities have been described (Ross et al.,
1970; Strott et al., 1970; Sherman and Korenman, 1974a; Ross, 1976).
Concerning plasma gonadotropins, most of the features observed in
normal cycles were present. However, the midcycle peaks of FSH and
LH were not always coincident, i.e., the FSH maximum occurred 1 day
after that of LH. As a whole, the mean FSH levels were lower than
those in control cycles and there was no increase at the end of the
luteal phase. LH concentrations tended to be higher before the midcycle peak and lower at the peak as well as during the luteal phase.
Moreover, FSH:LH ratios were decreased (Ross et al., 1970; Strott
et al., 1970; Ross, 1976). More recently, it has been shown that
in a high percentage of patients with short luteal phases, prolactin levels were slightly elevated (Mühlenstedt et al., 1977;
Seppälä et al., 1977; Del Pozo et al., 1979; Lehtovirta et al.,
1979).
        Concerning plasma steroids, 17-OHP plasma concentrations were
low in the follicular phase. Its pattern of gradual increase prior
to the LH peak was not observed, and its peak on the day of the
LH surge was, on the average, below the normal range and no secondary luteal rise could be demonstrated (Ross et al., 1970; Strott
et al., 1970; Sherman and Korenman, 1974a; Ross, 1976); the maximum
occurred much earlier than normal control cycles without reaching
5 ng/ml (Ross et al., 1970). Likewise, the area below the P curve
was markedly decreased compared with the controls (Del Pozo et al.,
1979).

        Alterations in 17β-estradiol levels, namely a low midcycle
peak and the absence of the expected rise during luteal phase,
were also demonstrated (Sherman and Korenman, 1974a). In the so-
called "inadequate luteal phase", a mild form of luteal deficiency,
hormonal alterations were reported to be less important. Plasma
P was reduced, FSH decreased during the follicular phase and at

the midcycle peak, while LH and E_2 were within the normal ranges (Sherman and Korenman, 1974b).

ETIOLOGY

Although ovarian factors may be responsible for some luteal phase defects (Jones, 1976), most cases seem to be due to a dysfunction of the hypothalamo-pituitary axis. Low preovulatory FSH levels could lead to abnormal follicular development and to inadequate luteinization or luteal function (Strott et al., 1970). This is substantiated by low levels of 17-OHP during the preovulatory phase and on the day of the LH peak, since 17-OHP is an index of follicular maturation (Strott et al., 1969). In addition, LH may play an etiological role in luteal insufficiency since its concentrations at the midcycle peak and during the luteal phase are lower than those in the normal menstrual cycle (Jones and Madrigal-Castro, 1970; Strott et al., 1970). Moreover, in vitro studies by McNatty and Sawers (1975) have demonstrated that the ability of human granulosa cells to undergo mitosis or to secrete P in culture is dependent on the hormonal environment of the follicle. Cells derived from follicles containing high concentrations of FSH, LH and E_2 secrete significantly more P than cells from follicles not containing all three hormones. Furthermore, luteal phase defects could be induced in anovulatory women given inadequate amounts of FSH (Jones et al., 1969), human chorionic gonadotropin (hCG; Jones et al., 1974), clomiphene (Jones and Madrigal-Castro, 1970; Van Hall and Mastboom, 1969), or LH (Vande Wiele et al., 1970).

Hyperprolactinemia has also been implicated as a causative factor in luteal deficiency according to many studies reviewed recently by Seppälä (1978) and Tourniaire (1979). Indeed, a high incidence of moderately elevated plasma prolactin levels has been found in women with short luteal phases (Mühlenstedt et al., 1977; Seppälä et al., 1977; Del Pozo et al., 1979; Lehtovirta et al., 1979), whereas according to Sarris et al. (1978), hyperprolactinemia could not be demonstrated in women exhibiting only repeated low P levels. It appears then that hyperprolactinemia is not found in all cases of luteal defect and even in the severe form, the anovulatory cycle, prolactin levels may be normal. However, in spite of normal basal prolactin levels, patients with anovulatory cycles may display prolactin hyperresponse to thyrotropin-releasing hormone (TRH; Peillon et al., 1979). Furthermore, in cycles with sulpiride-induced hyperprolactinemia, the LH surge either did not occur or was lower than in controls, and the ensuing luteal phase was short with reduced P levels (Delvoye et al., 1974). This in vivo inhibition of P production by hyperprolactinemia has also been demonstrated in vitro by McNatty et al. (1974) who have also shown that this inhibition is proportional to the prolactin dose added.

In the short luteal phase, bromocriptine treatment normalizes elevated prolactin levels and lengthens the luteal phase even in patients with initially normal prolactin concentrations (Mühlenstedt et al., 1977; Seppälä et al., 1977; Del Pozo et al., 1979; Lehtovirta et al., 1979). In anovulatory cycles, this therapy induces ovulation in most cases (Peillon et al., 1979). When the effect of bromocriptine on P production was examined by Seppälä et al. (1977) and Lehtovirta et al. (1979), no improvement was observed. Del Pozo et al. (1979), however, found P levels increased during this treatment, even though the dose (5 mg/day) used by these two groups was the same.

Oversuppression of prolactin may induce a decrease in P production in luteal-deficient patients (Mühlenstedt et al., 1977) as well as in normal women (Schulz et al., 1978). This agrees with in vitro data obtained by McNatty et al. (1974) who showed that neutralization of prolactin in culture medium by the addition of rabbit anti-human prolactin serum resulted in a significant decrease in P production by granulosa cells compared with controls. Thus, in vivo as well as in vitro studies demonstrate that small amounts of prolactin are necessary for normal corpus luteum function; however, high concentrations may induce short luteal phase. More recently, Gautray et al. (1978) suggested another approach to the etiology by studying electroencephalographic patterns and applying a therapy of nonhormonal drugs acting on the central nervous system.

Hyperandrogenism may be implicated in shortening the luteal phase. Smith et al. (1979) have demonstrated a correlation between a shortened luteal phase and testosterone levels in a group of infertile women with ovulatory cycles. Reduction of testosterone levels by chronic prednisone administration was associated with significant shortening of the follicular phase and lengthening of the luteal phase (Rodriguez-Rigau et al., 1979). Elevated androgen levels could be responsible for a dysfunction of the hypothalamo-pituitary axis resulting in inadequate follicular maturation or an inadequate LH surge. It should be noted, however, that plasma P was not determined in either of the above mentioned studies, so that corpus luteum function could not be evaluated in the presence of elevated testosterone levels or after prednisone testosterone suppression. Moreover, since Sowers et al. (1979) have shown that dexamethasone inhibits LH and FSH response to clomiphene, it would be interesting to assess the pituitary gonadotropins in the course of prednisone therapy.

Stimulation tests by hCG or clomiphene can be useful in diagnosing and subsequently correcting the disorder. hCG administration may determine not only the responsiveness of the corpus luteum to its physiologic stimulating factor in early pregnancy, but also its adequacy. Indeed, when a normal corpus luteum is

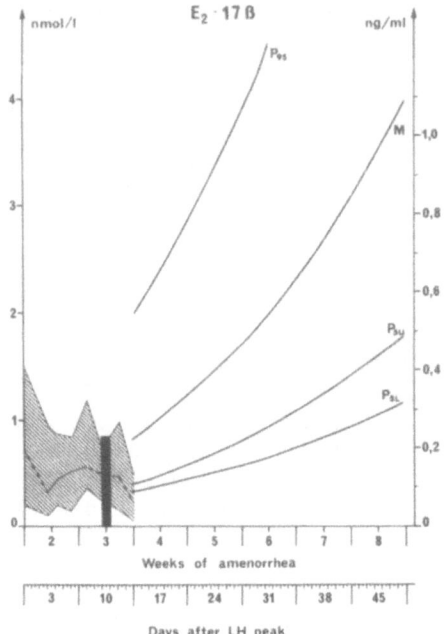
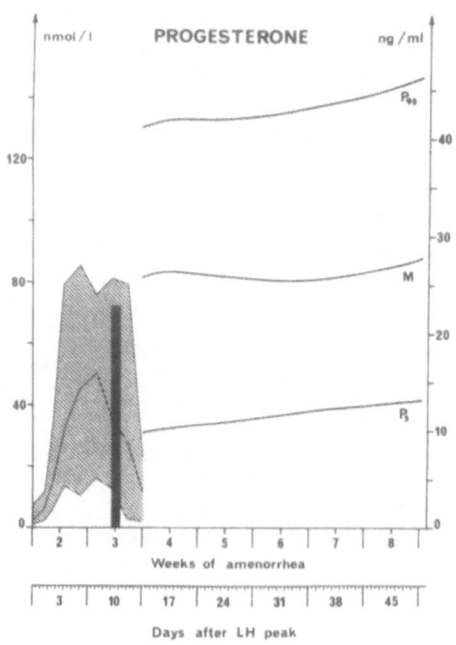

Fig. 8.   Normal plasma E_2 and P responses to hCG test.   Comparison
with levels observed in normal luteal phase and early
pregnancy.

stimulated by three intramuscular injections of 5000 IU of hCG on
the 3rd, 5th and 7th days of the thermal plateau, P and E_2 increase
and generally reach levels observed in early pregnancy (Fig. 8).

An impaired steroid response to hCG demonstrates a corpus
luteum dysfunction either due to a primary ovarian defect or to an
abnormality in follicle growth secondary to inadequate stimulation
by FSH.   This refractoriness of corpus luteum to hCG has been
suggested to be involved in recurrent early abortions (Jones et al.,
1974).   Dissociated responses of plasma P and E_2 have not yet been
observed in our practice, as reported for urinary metabolites (see
review Dubecq et al., 1980).   A positive response would incriminate
deficient corpus luteum stimulation by LH and therapy with hCG
would be indicated.

A clomiphene test, performed according to Geller's (1980)
protocol (daily administration of 250 mg of clomiphene from the
5th day to the 8th day of the thermal plateau), might provide more

information on the hypothalamo-pituitary-ovarian axis. A positive
response of ovarian steroids would indicate the integrity of the
whole axis. Isolated steroid unresponsiveness would suggest a
primary corpus luteum defect, whereas an impaired response by both
LH and steroids would imply a central origin.

CONCLUSION

Luteal phase defects suspected on the basis of the BBT chart
and endometrial biopsy must be confirmed by evaluation of P production. The most reliable index of this production seems to be
the area below the P curve which can be determined by the three P
levels obtained in blood samples collected between days M-11 and
M-4. This investigation should be completed by E_2 measurement,
since E_2 production is impaired when luteal insufficiency becomes
severe. Furthermore, since luteal defects may have multiple
etiologies, determinations of plasma androgens, gonadotropins and
prolactin should be performed. Thyrotropin- and luteinizing-
releasing hormone stimulation tests would detect any abnormality
in gonadotropin release or a further prolactin hyperresponse.
Stimulation by hCG and clomiphene may provide a therapeutic
orientation.

REFERENCES

Abraham, G.E., Maroulis, G.B., and Marshall, J.R., 1974, Evaluation
    of ovulation and corpus luteum function using measurements of
    plasma progesterone, Obstet. Gynecol., 44:522.
Andrews, W.C., 1979, Luteal phase defects, Fertil. Steril., 32:501.
Annos, T., Thompson, I.E., and Taymor, M.L., 1980, Luteal phase
    deficiency and infertility: difficulties encountered in
    diagnosis and treatment, Obstet. Gynecol., 5:705.
Baird, D.T., 1976, Ovarian steroid secretion and metabolism, in:
    "The Endocrine Function of the Human Ovary," V.H.T. James,
    M. Serio and G. Giusti, ed., New York, p. 125.
Cooke, K.D., Morgan, C.A., and Parry, T.E., 1972, Correlation of
    endometrial biopsy and plasma progesterone levels in infer-
    tile women, J. Obstet. Gynaecol. Br. Commonw., 79:647.
Crépin, G., Delcroi, M., Querleu, D., Scholler, R., and Castanier,
    M., 1980, Etude longitudinale de la grossesse débutante nor-
    male et pathologique au moyen des dosages hormonaux plas-
    matiques, J. Gynéc. Obstét. Biol. Reprod., 9:73.
Csapo, A.I., and Pulkkinen, M., 1978, Indispensability of the human
    corpus luteum in the maintenance of early pregnancy luteec-
    tomy evidence, Obstet. Gynecol. Surv., 33:69.
Csapo, A.I., Pulkkinen, M.O., and Kaihola, H.L., 1973a, The effect
    of estradiol replacement therapy on early pregnant luteec-
    tomized patients, Am. J. Obstet. Gynecol., 117:987.

Csapo, A.I., Pulkkinen, M.O., Ruttner, B., Sauvage, J.P., and
    Wiest, W.G., 1972, The significance of the human corpus
    luteum in pregnancy maintenance, I. Preliminary studies,
    Am. J. Obstet. Gynecol., 112:1061.

Csapo, A.I., Pulkkinen, M.O., and Wiest, W.G., 1973b, Effects of
    luteectomy and progesterone replacement therapy in early
    pregnant patients, Am. J. Obstet. Gynecol., 115:759.

Csapo, A.I., and Resch, B., 1979, Prevention of implantation by
    antiprogesterone, J. Steroid Biochem., 11:963.

Del Pozo, E., Wyss, H., Tolis, G. Alcaniz, J., Campana, A., and
    Naftolin, F., 1979, Prolactin and deficient luteal function,
    Obstet. Gynecol., 53:282.

Delvoye, P., Taubert, H.-D., Jürgensen, O., L'Hermite, M., Delogne,
    J., and Robyn, C., 1974, Influence of circulating prolactin
    increased by a psychotropic drug on gonadotropin and proges-
    terone secretion, Acta Endocrinol. [Suppl.] (Kbh) 184, Abstr.
    110,110.

Dhont, M., Vandekerckove, D., Vermeule, A., and Vanderweghe, M.,
    1974, Daily concentrations of plasma LH, FSH, estradiol,
    estrone and progesterone throughout the menstrual cycle,
    Eur. J. Obstet. Gynec. Reprod. Biol. [Suppl. 1] 4:S153.

Dubecq, J.P., and Gonnet, J.M., 1980, II. Prevention des avorte-
    ments endocriniens par stimulations du follicule, J. Gynecol.
    Biol. Reprod., 9:95.

Gambrell, R.D., 1979, The role of hormones in the etiology of
    breast and endometrial cancer, Acta Obstet. Gyne. Scand.
    [Suppl.] 88:73.

Gautray, J.P., De Brux, J., Azadian-Boulanger, G., Jolivet, A.,
    Colin, M.C., and Margenet-Baudry, M., 1978, Physiologie et
    physiopathologie du cycle menstruel humain à l'époque de
    l'implantation. Aspects hormonaux et endométriaux, in:
    "L'implantation de l'oeuf," F. Du Mesnil Du Buisson, A.
    Psychoyos, and K. Thomas, ed., Masson, Paris, p. 35.

Gautray, J.P., Jolivet, A., Goldenberg, A., Tajchner, G., and
    Eberhard, A., 1978, Clinical investigation of the menstrual
    cycle. II. Neuroendocrine investigation and therapy of the
    inadequate luteal phase, Fertil. Steril., 29:275.

Geller, S., 1961, "La courbe thermique," Masson, Paris.

Geller, S., 1980, Personal communication.

Geller, S., Grenier, J., Nahoul, K., and Scholler, R., 1979,
    Insuffisance lutéale et mastopathies bénignes. Etude à la
    lumière des données de l'épreuve combinée LH-RH + TRH
    couplée à l'étude des stéroides ovariens, Ann. Endocrinol.,
    40:45.

Hensleigh, P.A., and Fainstat, T., 1979, Corpus luteum dysfunction:
    serum progesterone levels in diagnosis and assessment of
    therapy for recurrent and threatened abortion, Fertil. Steril.,
    32:396.

Hilgers, T.W., and Bailey, A.J., 1980, Natural family planning.
II. Basal body temperature and estimated time of ovulation,
Obstet. Gynecol., 32:333.

Horta, J.L.H., Fernandez, J.G., de Leon, B.S., and Cortes-Gallegos,
V., 1977, Direct evidence of luteal insufficiency in women
with habitual abortion, Obstet. Gynecol., 49:705.

Israel, R., Mishell, D.R., Stone, S.C., Thorneycroft, I.H., and
Moyer, D.L., 1972, Single luteal phase serum progesterone
assay as an indicator of ovulation, Am. J. Obstet. Gynecol.,
112:1043.

Johansson, E.D.B., 1969, Progesterone levels in peripheral plasma
during the luteal phase of the normal human menstrual cycle
measured by a rapid competitive protein binding technique,
Acta. Endocrinol., (Kbh), 61:592.

Johansson, E.D.B., Larsson-Cohn, U., and Gemzell, C., 1972, Monophasic basal body temperature in ovulatory menstrual cycles,
Am. J. Obstet. Gynecol., 113:933.

Jones, G.E.S., 1949, Some newer aspects of the management of infertility, J. Am. Med. Assoc., 114:1123.

Jones, G.S., 1975, Luteal defects, in: "Progress in Infertility,"
S.J. Behrman and R.W. Kistner, ed., R.W. Churchill,
London, p. 299.

Jones, G.S., 1976, The luteal phase defect, Fertil. Steril., 27:
351.

Jones, G.S., Aksel, S., and Wentz, A.C., 1974, Serum progesterone
values in the luteal phase defects. Effect of chorionic
gonadotropin, Obstet. Gynecol., 44:26.

Jones, G.S., Ruehsen, M.D.M., Johanson, A.J., Raiti, S., and
Blizzard, R.M., 1969, Elucidation of normal ovarian physiology by exogenous gonadotropin stimulation following
steroid pituitary suppression, Fertil. Steril., 20:14.

Jones, G.S., and Madrigal-Castro, V., 1970, Hormonal findings in
association with abnormal corpus luteum function in the
human: the luteal phase defect, Fertil. Steril., 21:1.

Keller, D.W., Wiest, W.G., Askin, F.B., Johnson, L.W., and
Strickler, R.C., 1979, Pseudocorpus luteum insufficiency:
a local defect of progesterone action on endometrial stroma,
J. Clin. Endocrinol. Metab., 48:127.

Kletsky, O.A., Nakamura, R.M., Thorneycroft, I.H., and Mishell,
D.R., 1975, Log normal distribution of gonadotropins and
ovarian steroid values in normal menstrual cycle, Am. J.
Obstet. Gynecol., 121:688.

Lehtovirta, P., Arjomaa, P., Ranta, T., Laatikainen, T., Hirvonen,
E., and Seppälä, M., 1979, Prolactin levels and bromocriptine
treatment of short luteal phase, Int. J. Fertil., 24:57.

Mauvais-Jarvis, P., Ware, R., Sterkers, N., Ohlgiesser, C., and
Tamborini, A., 1976, Corps jaune inadéquat et mastopathies,
Ann. Endocrinol., 37:347.

McNatty, K.P., and Sawers, R.S., 1975, Relationship between the
    endocrine environment within the Graafian follicle and the
    subsequent rate of progesterone secretion by human granulosa
    cells in vitro, J. Endocrinol., 66:391.

McNatty, K.P., Sawers, R.S., and McNeilly, A.S., 1974, A possible
    role for prolactin in control of steroid secretion by the
    human Graafian follicle, Nature, 250:653.

Moghissi, K.S., 1976, Accuracy of basal body temperature for ovu-
    lation detection, Fertil. Steril., 27:1415.

Mühlenstedt, D., Wuttke, W., and Schneider, H.P.G., 1977, Short
    luteal phase and prolactin, Fertil. Steril., 28:373.

Murthy, Y.S., Arronet, G.H., and Parekh, M.C., 1970, Luteal phase
    inadequacy. Its significance in infertility, Obstet. Gynecol.,
    36:758.

Noyes, R.W., Hertig, A.T., and Rock, J., 1950, Dating the endome-
    trial biopsy, Fertil. Steril., 1:3.

Peillon, F., Vincens, M., Cesselin, F., Doumith R., and Mowszowicz,
    I., 1979, Cycles anovulatoires avec normoprolactémie apparente.
    Traitement par bromocriptine, Nouv. Presse Med., 8:3269.

Quagliarello, J., and Weiss, G., 1979, Clomiphene citrate in the
    management of infertility associated with shortened luteal
    phases, Fertil. Steril., 31:373.

Radwanska, E., and Swyer, G.I.M., 1974, Plasma progesterone esti-
    mation in infertile women and in women under treatment with
    clomiphene and chorionic gonadotrophin, J. Obstet. Gynaec.
    Br. Commonw., 81:107.

Reed, A.H., Henry, R.J., and Mason, W.B., 1971, Influence of
    statistical method used on the resulting estimate of normal
    range, Clin. Chem., 17:275.

Rodriguez-Rigau, L.J., Smith, K.D., Tcholakian, R.K., and Stein-
    berger, E., 1979, Effect of prednisone on plasma testosterone
    levels and on duration of phases of the menstrual cycle in
    hyperandrogenic women, Fertil. Steril., 32:408.

Rosenberg, S.M., Luciano, A.A., and Riddick, D.H., 1980, The luteal
    phase defect: the relative frequency of, and encouraging
    response to, treatment with vaginal progesterone, Fertil.
    Steril., 34:17.

Rosenfeld, D.L., and Garcia, C.-R., 1976, A comparison of endome-
    trial histology with simultaneous plasma progesterone deter-
    minations in infertile women, Fertil. Steril., 27:1256.

Ross, G.T., 1976, Preovulatory determinants of human corpus luteum
    function, Eur. J. Obstet. Gynecol. Reprod. Biol., 64:147.

Ross, G.T., Cargille, C.M., Lipsett, M.B., Rayford, P.L., Marshall,
    J.R., Strott, C.A., and Rodbard, D., 1970, Pituitary and
    gonadal hormones in women during spontaneous and induced
    ovulatory cycles, Recent Prog. Horm. Res., 26:1.

Sarris, S., Swyer, G.I.M., McGarrigle, H.H.G., Lawrence, D.M.,
    Little, B., and Lachelin, G.C.L., 1978, Prolactin and
    luteal insufficiency, Clin. Endocrinol., 9:543.

Scholler, R., Auric, F., Albigès, C.H., and Del Pozo, D., 1974, Variabilité des taux de l'estriol urinaire au cours de la second moitié de la grossesse normale. III. Variabilité, inter-individuelle. Valeurs de référence. Corrélations, Revue Fr. Gynecol. Obstet., 69:429.

Scholler, R., Castanier, M., Feinstein, M.-C., Grenier, J., Nahoul, K., Roger, M., Abeille, J.-P., and Legros, R., 1978, Reference values during the menstrual cycle. Anatomical-biochemical correlations, in: "Endocrinology of the ovary," R. Scholler, ed., Proceedings of the International Symposium, Paris-Fresnes, SEPE, Paris, p. 257.

Scholler, R., Chartier, M., Barrat, J., Roger, M., Castanier, M., and Avigdor, R., 1980, D. Profil de hcG et des stéroides dans les avortements spontanés du 1er trimestre, J. Gynec. Obstet. Biol. Reprod., 9:79.

Scholler, R., Roger, M., and Avigdor, R., 1973a, Inter- and intra-individual variations in reference values for hormone assays, in: "Reference Values in Human Chemistry," G. Siest, ed., Karger, Basel, p. 47.

Scholler, R., Roger, M., Grenier, J., Nahoul, K., Castanier, M., and Adeline, J., 1977, Interprétation des dosages plasmatiques, in: "Actualités gynécologiques, 8ème série," A. Netter and A. Gorins, ed., Masson, Paris, p. 213.

Scholler, R., Soldat, M.-C., and Avigdor, R., 1973b, Elimination de normale. Corrélations avec le poids du nouveau-né. Variations journalieres, Pathol. Biol., 21:375.

Schulz, K.-D., Geiger, W., Del Pozo, E., and Künzig, H.J., 1978, Pattern of sexual steroids, prolactin, and gonadotropic hormones during prolactin inhibition in normally cycling women, Am. J. Obstet. Gynecol., 132:562.

Seppälä, M., 1978, Prolactin and female production, Ann. Clin. Res., 10:164.

Seppälä, M., Lehtovirta, P., Laatikainen, T., Ranta, T., Hirvonen, E., and Arjomaa, P., 1977, Bromocriptine treatment of luteal insufficiency, Acta Endocrinol. [Suppl.] (Kbh) 212:43.

Seppälä, M., Lehtovirta, P., and Rutanen, E.-M., 1978, Detection of chorionic gonadotrophin-like activity in infertile cycles with a short luteal phase, Acta Endocrinol., 88:164.

Shepard, M.K., and Senturia, Y.D., 1977, Comparison of serum progesterone and endometrial biopsy for confirmation of ovulation and evaluation of luteal function, Fertil. Steril., 28:541.

Sherman, B.M., and Korenman, S.G., 1974a, Measurement of plasma LH, FSH, estradiol and progesterone in disorders of the human menstrual cycle: the short luteal phase, J. Clin. Endocrinol. Metab., 38:89.

Sherman, B.M., and Korenman, S.G., 1974b, Measurement of serum LH, FSH, estradiol and progesterone in disorders of the human menstrual cycle: the inadequate luteal phase, J. Clin. Endocrinol. Metab., 39:149.

Sherman, B.M., and Korenman, S.G., 1974c, Inadequate corpus luteum function: a pathophysiological interpretation of human breast cancer epidemiology, Cancer, 33:1306.

Sitruk-Ware, R., Sterkers, N., and Mauvais-Jarvis, P., 1979, Benign breast disease I: hormonal investigation, Obstet. Gynecol., 53:457.

Smith, K.D., Rodriguez-Rigau, L.J., Tcholakian, R.K., and Steinberger, E., 1979, The relationship between plasma testosterone levels and the lengths of phases of the menstrual cycle, Fertil. Steril., 32:403.

Soules, M.R., Wiebe, R.H., Aksel, S., and Hammond, C.B., 1977, The diagnosis and therapy of luteal phase deficiency, Fertil. Steril., 28:1033.

Sowers, J.R., Rice, B.F., and Blanchard, S., 1979, Effect of dexamethasone on luteinizing hormone and follicle stimulating hormone responses to LHRH and to clomiphene in the follicular phase of women with normal menstrual cycles, Horm. Metab. Res., 11:478.

Strott, C.A., Cargille, C.M., Ross, G.T., and Lipsett, M.B., 1970, The short luteal phase, J. Clin. Endocrinol. Metab., 30:246.

Strott, C.A., Yoshimi, T., Ross, G.T., and Lipsett, M.B., 1969, Ovarian physiology: relationship between plasma LH and steroidogenesis by the follicle and corpus luteum; effect of HCG, J. Clin. Endocrinol. Metab., 29:1157.

Tourniaire, J., 1979, Hyperprolactinémie et insuffisance lutéale, Ann. Endocrinol., 40:342.

Tredway, D.R., Mishell, D.R., and Moyer, D.L., 1973, Correlation of endometrial dating with luteinizing hormone peak, Am. J. Obstet. Gynecol., 117:1030.

Vande Wiele, R.L., Bogumil, J., Dyenfurth, I., Férin, M., Jewelewicz, R., Warren, M., Rizkallah, T., and Mikhail, G., 1970, Mechanisms regulating the menstrual cycle in women, Recent Prog. Horm. Res., 26:63.

Van Hall, E.V., and Mastboom, J.L., 1969, Luteal phase insufficiency in patients treated with clomiphene, Am. J. Obstet. Gynecol., 103:165.

Venning, E.M., and Henry, J.S., 1937, The measurement of a pregnandiol complex in human urine, Can. Med. Assoc. J., 83.

Yip, S.K., and Sung, M.L., 1977, Plasma progesterone in women with a history of recurrent early abortions, Fertil. Steril., 28:151.

EVALUATION OF OVARIAN DISTURBANCES BY ENDOMETRIAL BIOPSY

Jean de Brux

Institut de Pathologie et de Cytologie Apliquée
53, rue des Belles-Feuilles
75116 Paris, France

Confronted with a cyclic disturbance and/or a problem of infertility, the gynecologist resorts to endometrial biopsy to eliminate simple or specific inflammation, dystrophy, or atypical adenomatous hyperplasia, and to determine whether hormonal disturbances are involved; these may be confirmed by steroid or gonadotropin concentrations. If endometrial responses appear to be incomplete, it is often because of the imprecision or insufficiency of clinical and biological data furnished to the pathologist.

MATERIALS AND METHODS

Biopsies, immersed in an adequate fixative, such as Bouin or Carnoy (formalin and alcohol must be excluded), were taken between days 20 and 24, the presumed period of possible implantation. Microscopic examination (329 biopsies) was carried out without clinical information by the same pathologist, to ensure uniform interpretation. Clinical and biological discussions took place regularly between the clinician (J.P. Gautray) and the biochemist (P. Robel).

The average patient age was 29.5 years. The reasons for the first examination were: cyclic anomaly (93 cases); pelvic pain (25 cases); amenorrhea (17 cases); primary sterility (82 cases); and secondary sterility (112 cases). It was sometimes necessary to add a PAS stain and reticulin to classic staining by hematoxylin-eosin-safran.

To correctly evaluate a hormonal disturbance, it is essential to know the effects of hormones on the endometrium during the course of the normal 28 day cycle (see Noyes et al., 1950; Philippe et al.,

107

1965; Richart and Ferenczy 1974; Rosenfeld and Ramon-Garcia 1975).

NORMAL MENSTRUAL CYCLE

   The menstrual cycle begins with the first day of menstruation.
From this moment on, follicles, stimulated since the previous ovulation, undergo a much stronger response to FSH, and secrete estrogens in greater quantity, assuring the proliferative phase of the
endometrium. On day 14, ovulation occurs, induced by a peak of LH
followed immediately by a peak of FSH. The action of LH determines
luteal transformation of the follicle and progesterone secretion
which may reach 25 ng/24 h. It must be noted that the luteinized
internal theca continues to secrete estrogens, but in lesser
amounts than the follicle.

   The normal endometrial cycle may be divided into three princi-
pal phases: (1) the follicular phase, in which three periods may
be distinguished, early (days 5 to 7), middle (days 8 to 10), and
late (days 11 to 14); (2) the secretory phase, which also has three
periods, early (days 15 to 18), middle (days 19 to 23) and late
(days 24 to 28); and (3) the menstrual phase (days 1 to 4).

   The follicular (or proliferative) phase is characterized by
endometrial growth. In the early period, the endometrium is only
1 to 2 mm thick. The epithelial cells are cuboid in shape, sometimes with a few lipid inclusions, glycogen is not secreted, and
mitoses are rare. Invaginations of the surface epithelium form
straight, glandular tubes.

   During the middle follicular phase, the tubes multiply and
increase in length, cell mitosis increases, with frequent multi-
nucleation resulting in pseudostratification, and mucoid or glycogen secretion is not evident. The stroma, which in the previous
phase was compact and filled with elongated mesenchymal cells,
becomes edematous in the middle follicular phase. The cell cytoplasm becomes larger and clearly outlined, mitoses increase and
capillaries become numerous.

   In the late follicular phase, the endometrial mucosa assumes
an undulent surface and the glands become dilated and twisted.
The border cells show clear pseudostratification, and at the apical
part, sometimes there is slight polysaccharide vacuolization,
especially if estrogen stimulation has been strong. Immediately
preceding the preovulatory period, the presence of a few small
glycogenic basal vacuoles indicates the beginning of luteal action.
This is detectable chiefly by electron microscopy. Mitoses persist
in the stroma, but are less abundant, and edema diminishes.

   The early period of the secretory (or luteal or folliculo-
luteinic) phase (days 15 to 18) is characterized by the presence

of clear subnuclear vacuoles, which appear 8 to 36 hours after
ovulation, depending on corpus luteum activity. At the same time,
mitoses disappear and the lumen of the tubes widens. On days 17
and 18, glycogen vacuoles in the glandular epithelium rise to the
apical part of the cells crowding the nuclei.

On day 19 (middle secretory phase), the nuclei return to the
base of the cells and excretion of glycogen into the lumen begins
on day 20. The substance is eosinophilic and PAS-positive. The
period from days 20 to 23 is characterized by stromal activity.
Edema, which is constant, suddenly increases on day 21, and is
maximal on day 23. The cells are small and crowded together and
have only a thin circle of cytoplasm. Growth of the cytoplasm
determines the appearance of: (1) the connective spines, which form
the "saw-teeth" of the tubes and increase the surface area for better
nutrition of the ovum if implantation occurs; and (2) the spiral
arteries, which are well formed on day 23 of the cycle. The spiral
arteries relate to the action of progesterone on the capillaries.
The pericytes widen and the myofibrils become differentiated. The
stromal cells widen, form a cuff and undergo eosinophilic trans-
formations, thus becoming the first predecidual cells.

In the late luteal secretory phase, these phenomena are inten-
sified. The connective spines continue to grow and the lumen of
the tubes narrows. The border cells decrease in length, the nuclei
retract and the apical part of the cells herniates into the lumen,
similar to apocrine cells during secretion (Wynn and Woolley, 1967;
Parr and Parr, 1974). These cells seem to be composed of neutral
mucopolysaccharides, lipids and water. The spiral arteries become
longer and thicker, and, upon section, form nests surrounded by an
extensive zone of decidual cells that appear beneath the surface
epithelium. By days 26 and 27, infiltration of polynuclears and
lymphocytes is observed, and larger cells, the "kornzellen" (Hamperl
1954; Dallenbach-Hellweg 1975) secrete relaxin. Some intratissue
bleeding due to vasodilation and a vasoparalysis of the capillaries
is noted; these zones rapidly become small necrotic foci.

Thus, at the end of the secretory phase, three distinct layers
may be found in the endometrium: (1) the compact zone, distinguished
by the presence of predecidualized stromal cells; (2) the spongy
zone, characterized by numbers of dilated and twisted tubes; and
(3) the basal zone or regenerative area, in contact with the myometrium and formed by tubes surrounded by dense stroma. It is from
this layer and the remains of the lower third of the spongy zone
that the endometrium is reconstituted after abrasion.

IMPACT OF HORMONAL DISTURBANCES ON THE ENDOMETRIUM

Even if the responses of the various constituents of the

endometrium to hormonal stimuli enable the histologist to give the
clinician a precise picture of ovarian activities (usually confirmed
by plasma concentrations), some discrepancies may result because of
clinical information, the dates of respective biopsies and the
presumed moment of ovulation.

Estrogenic Insufficiency

A total absence of estrogenic stimulation results in endometrial
atrophy, the presence of shreds of cylindrical surface epithelia or
fragments of endometrium limited to the zone of regeneration, with
few glandular tubes (some flattened, others slightly dilated) in a
stroma composed of crowded fibrocytes. The nuclei are retracted
and the chromatin is blotted out. This is typical of the atrophic
postmenopausal endometrium.

If estrogen deficiency is not total, the endometrium appears
hypotrophic, the stroma shows a slight interstitial edema, the
tubes are more abundant, with nuclei having denser and more evenly
distributed chromatin, visible nucleoli, and with cytoplasm that
is spread out. The epithelium of the tubes and of the surface
shows few mitoses.

Various etiologies result in such clinical pictures. In
young women presenting primary amenorrhea, an atrophic endometrium
indicates a total absence of estrogen and a very severe hypophyseal
ovarian disturbance, which must be clearly defined. A very hypotrophic endometrium showing a few mitoses indicates that the
follicles are poorly stimulated, and that the disturbance is of
hypothalamo-hypophyseal origin (Fig. 1). This condition is also
found in oligomenorrheic patients, in women with secondary
amenorrhea occurring either spontaneously or after interruption
of progestative treatments, or in instances of polycystic ovaries,
psychologic disturbances, prolactin adenoma, and suprarenal hyperplasia.

During perimenopause or menopause, the situation is somewhat
different. A profound hypophyseal-ovarian insufficiency may exist,
that may be transitory, with late or episodic reappearances of a
normal cycle. In established menopause, a subtrophic or hypotrophic
endometrium indicates pathologic resumption of estrogenic stimulation
because of an ovarian tumor with endocrine activity; estrogen transformation of androstenedione secreted by ovarian stroma or the
suprarenal; estrogen treatment for menopause, rheumatism or cosmetic
reasons; or administration of estrogen-like substances, such as
digitalin derivatives.

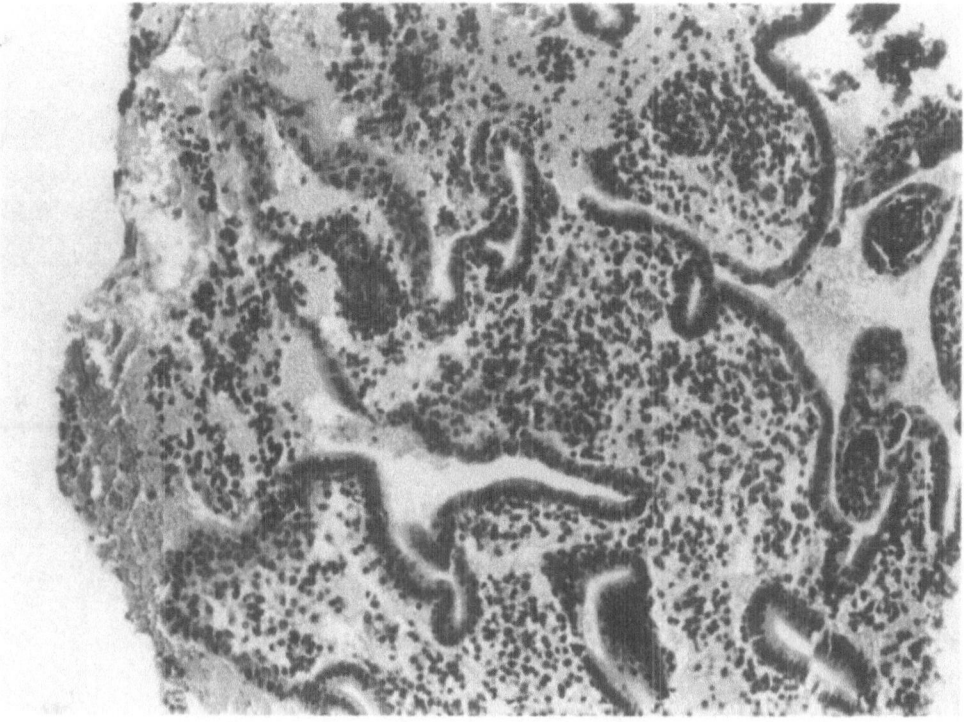

Fig. 1. Hypoplastic endometrium induced by weak stimulus
(chromatic nuclei, very few mitoses).

Luteal Insufficiency

Luteal insufficiency is common, but is often difficult to
diagnose based on endometrial biopsy since the anomalies are
slight. However, these could lead to infertility or to early
abortions (Horta et al., 1977; Seegar-Jones, 1977).

Luteal insufficiency associated with estrogen insufficiency
is manifested by only minor signs. The thinness of the mucosa
is easily recognized, since basal endometrium and even some fragments of myometrium are found on biopsy. Moreover, even if all
the constituents appear normal, the tubes are sparse and barely
wavy, connective spines are rare and barely visible, and glycogen
secretion is low. The stromal cells are narrow with clear nuclei.
The spiral arteries are rare and poorly differentiated (Fig. 2).
This clinical picture indicates a subnormal level of plasma steroids;
these endometria may accept nidation, but are unable to maintain it.

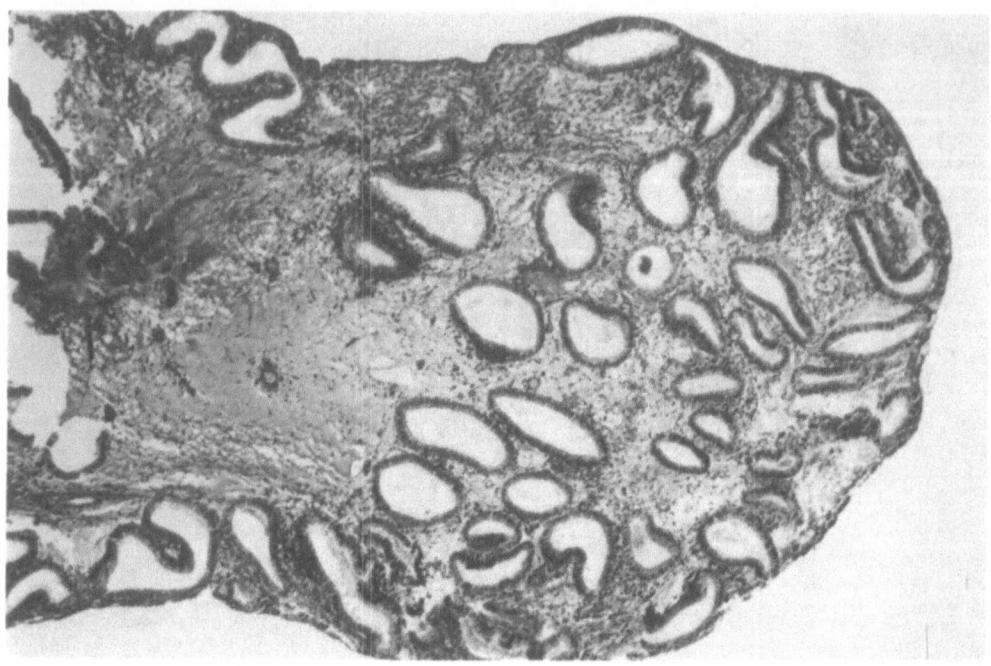

Fig. 2.   Ovarian insufficiency. The stroma is loose, glandular
          tubes are few and overstretched, with poor secretion and
          no spiral arteries.

      In luteal delay and/or deficiency without estrogen predom-
inance, the cycle is normal or sometimes slightly prolonged, and
the endometrium appears secretory (Fig. 3). However, closer
study of biopsies from days 22 and 23 shows greater than normal
edematous stroma filled with stellate or elongated cells with
narrow cytoplasm, and only slightly chromatic nuclei. The tubes
are dilated and wavy, but secretion is poor and delayed, and often
situated at the basal part of the cells, as in the secretory stage
(days 17 to 20 of the cycle). Spiral arteries are rare and poorly
differentiated. The level of plasma progesterone, though low, is
often barely subnormal. Biopsy taken at days 25 to 28 of the
cycle resembles that of days 21 to 22. It might be expected that
the endometrium would be hormonally capable of nidation, but the
survival period of the blastocyst has already terminated.

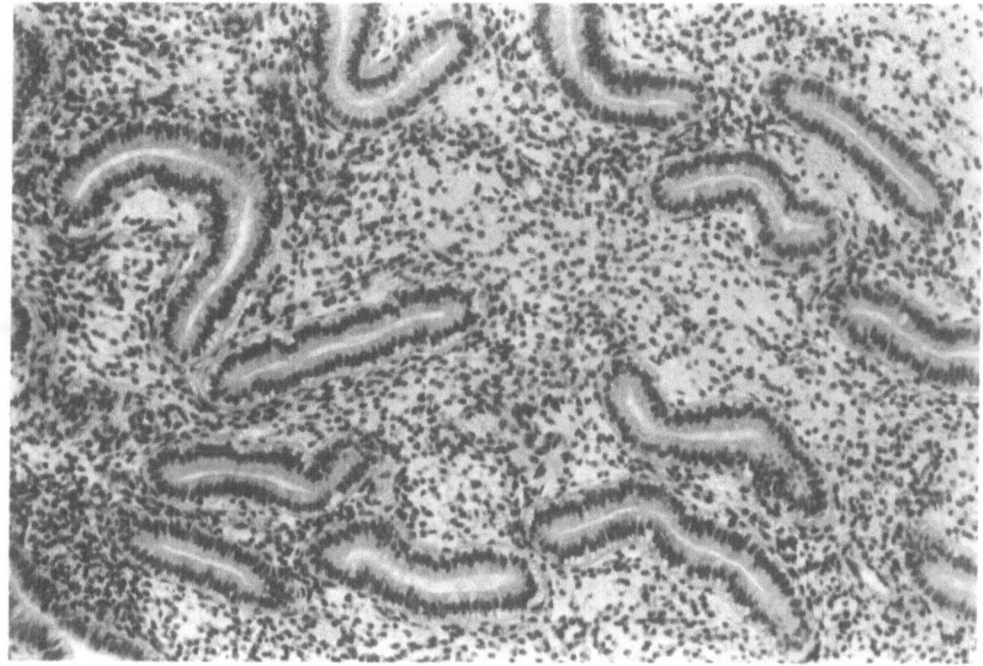

Fig. 3.   Luteal insufficiency with delayed glycogen secretion
          without estrogen predominance. Glandular tubes are
          regular but slightly wavy. Note poor glycogen secretion
          at the base of the cell and few differentiating spiral
          arteries. This biopsy taken on day 23 has the appearance
          of an endometrium on day 16.

     Luteal insufficiencies with estrogen predominance are common,
easily detected and constitute the so-called "endometrial dysmaturation" (Fig. 4). The endometrium is usually thick, and the stroma
is dense and rich in fibrocytes and fibroblasts, but lacks deposits
of collagen fibers. The glandular tubes are relatively numerous,
generally somewhat dilated, and regular or slightly wavy. The
cells of the lining epithelium show dense chromatin, sometimes
with multistratification and numerous mitoses. Glycogen secretion
may be either basal or apical. The spiral arteries are thick
walled, and appear well differentiated and grouped. These features
are characteristic of the endometrium on days 15 to 16; however,
the presence of the spiral arteries correctly correspond to the

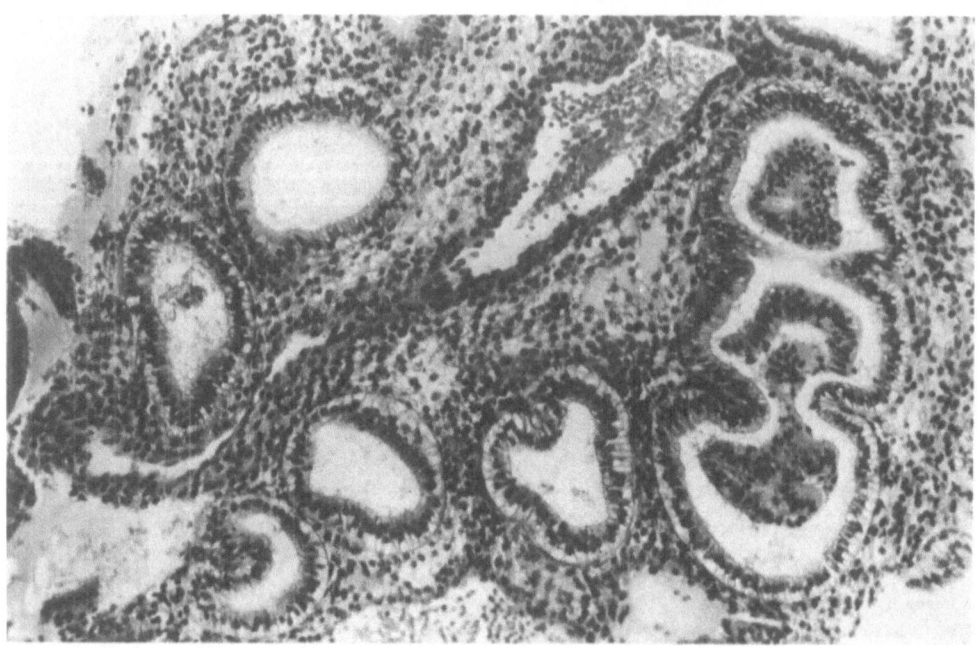

Fig. 4. Insufficient and delayed luteal response with estrogen
       predominance. Glandular tubes are overstretched, mitoses
       are numerous, and the epithelium has lucid zones. The
       stroma cells are fibroblasts.

real day of the cycle, days 25 to 28. Such endometria are often
found in sterile women.

     These immature endometria are characterized by marked fibro-
blastic hyperplasia of the stroma, and numerous wavy tubes crushed
by stromal proliferation. The stromal cells show an apical or
perinuclear secretion containing a few granules of glycogen and
are sometimes eosinophilic. The nuclei are irregular, notched and
bilobed, with few mitoses. There is also diffuse, subacute,
chronic endometritis. These anomalies represent minor Arias-
Stella atypias, and this form of endometrial dysmaturation is a
sequela of previous interruption of pregnancy. FSH stimulation,
reappearing after interruption of pregnancy, leads to permanent
estrogen secretion, and the chorion, persisting within the uterine
cavity, blocks LH secretion.

Diagnosis of premature arrest of the corpus luteum (Fig. 5) can be confirmed only by repeated determination of progesterone concentration; however, certain features of the endometrium permit pathological diagnosis. The mucosa has normal thickness due to eosinophilic edema of the stroma, sometimes with a few hemorrhages. The cells show some mitoses and are polymorphous: sometimes star-shaped with a poorly outlined cytoplasm, but often rounded and weakly eosinophilic, with edematous nuclei whose chromatin is pushed toward the periphery. The spiral arteries are often situated in the edema, and may be either well differentiated (sometimes resembling those of a hyperplastic endometrium because of the thickness of their walls) or poorly differentiated with

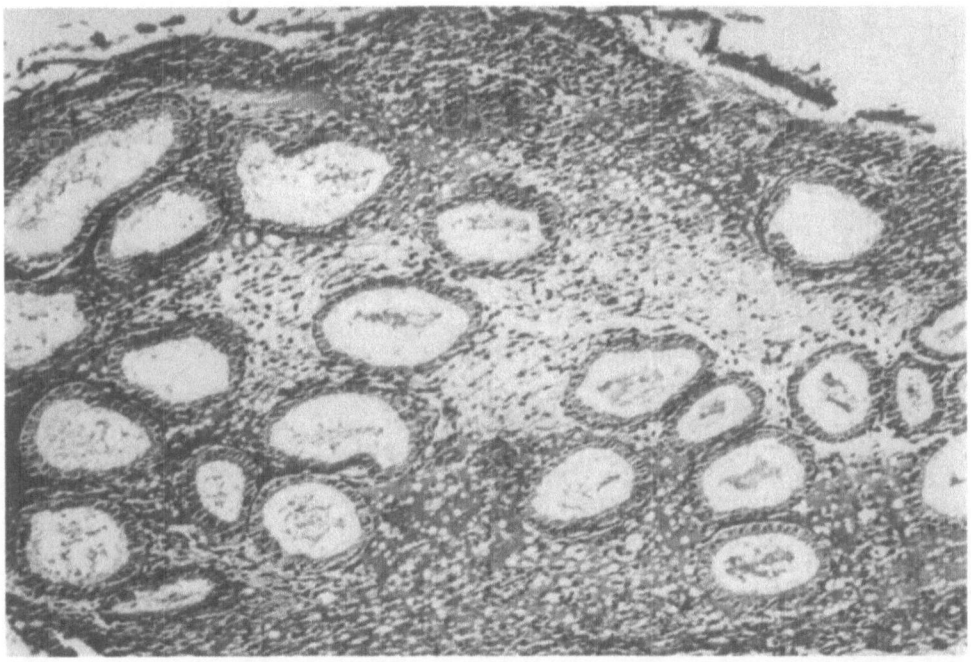

Fig. 5. Early corpus luteum deficiency. Glandular tubes are
         overstretched, with poor glycogen secretion. There are
         few spiral arteries. The stromal ground substance is
         edematous and fibrous.

poorly structured walls. The tubes are sparse but distended,
sometimes almost cystic and regular to slightly wavy. The epithelia of the tubes are hypotrophic, with clear-centered nuclei
similar to those of the stromal cells, and swollen, clear cytoplasm cut off at the apex, containing small secretory vacuoles
in the process of expulsion, as at the end of a cycle.

A study of each component of the endometrium sometimes reveals
discrepancies between the responses of the tubes and those of
the stroma during the secretory phase. These differences may be
found in the endometria of women presenting sterility of unex-
plained etiology. Hachiya et al. (1976) and Keller et al. (1979)
have made approximately identical observations. Such discrepancies
are linked to the premature response or lack of response of the
endometrial cells to progesterone. By days 23 to 24 (Fig. 6),
the cells have not widened, whereas the tubes show normal progesterone influence. There is an excessive decidualization in the

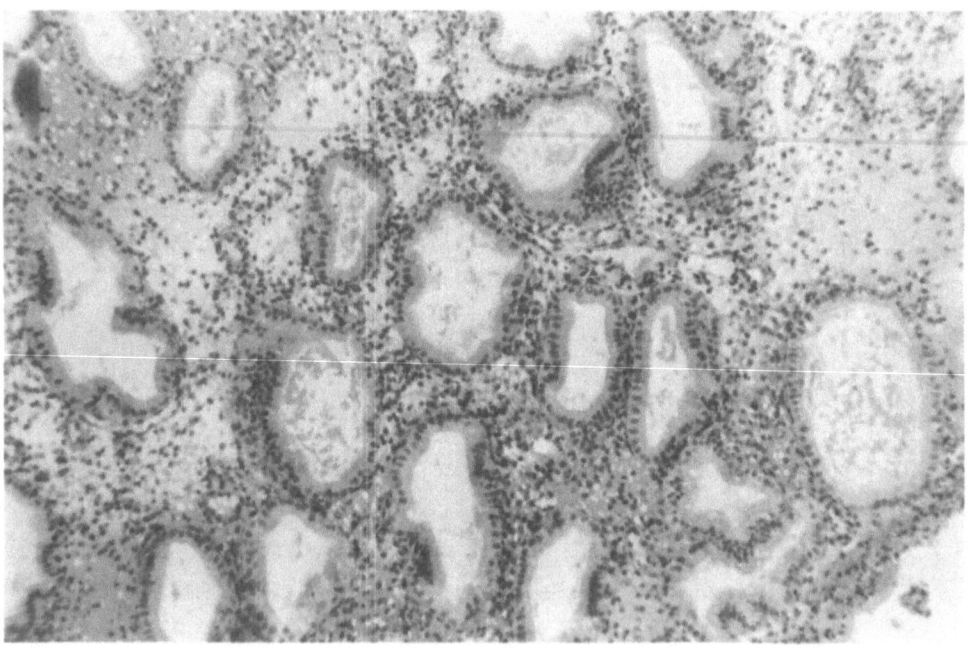

Fig. 6.   Discrepancy of the endometrial components under hormonal
          action. Endometrial biopsy on day 23. Note normal aspect
          of tubes. Endometrium is of ovulatory or postovulatory
          type.

subepithelial zone and/or around the spiral arteries; the tubes
are regular or barely wavy, with a weak glycogen charge situated
at the basal and apical parts of the cells, as on days 17 to 18
(Fig. 7).   The discrepancies between the responses of these two
constituents vary from 2 to 4 days; however, the true date of
the cycle may be determined by the state of the spiral arteries.

Hormonal Excesses (Excessive Estrogen Stimulation)

These differ according to whether the stimulation is isolated,
strong or prolonged or moderate and prolonged, or whether associated with a luteal influence, normal or weak, cyclic or irregular.   Isolated, strong and prolonged stimulation is easily
diagnosed and is characterized by diffuse polypoid hyperplasia of
the endometrium.   The stroma is dense, and the cells are elongated
or slightly rounded, and crowded together.   The tubes are numerous
and regular, and their epithelia are multistratified with numerous
mitoses.   This type of hyperplasia causes metrorrhagias due to

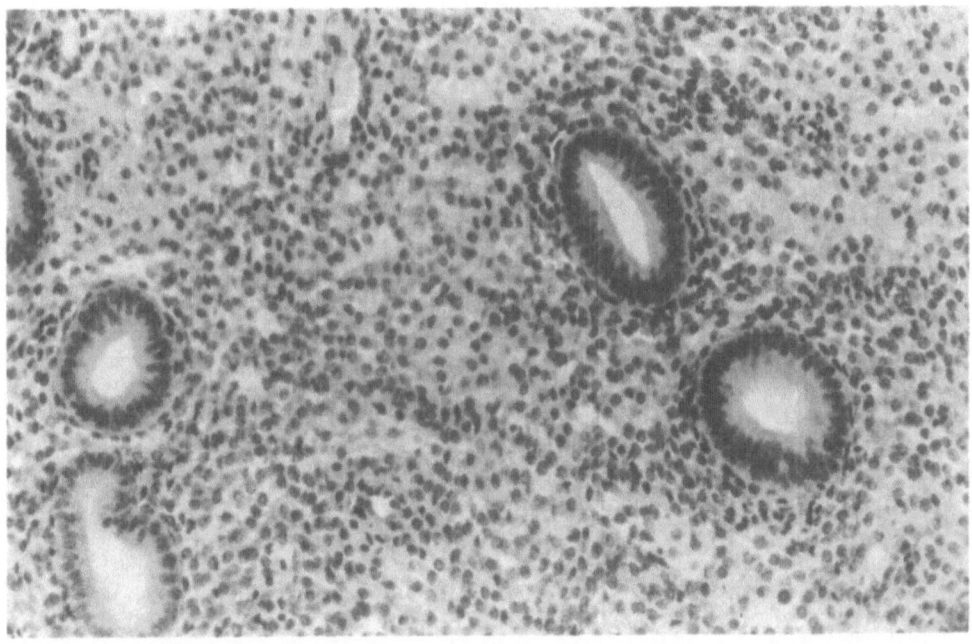

Fig. 7.   Discrepancy of the endometrial components under hormonal
          action.   Endometrial biopsy on day 23.   Note predecidual
          stroma and glandular tubes with postovulatory character-
          istics.

decreased hormonal stimulation. Hysterography shows a diffuse
lacunary appearance.

Isolated, moderate, and prolonged estrogenic stimulation
often goes unnoticed. The clinical symptom is amenorrhea or merely
a poly- or oligomenorrhea. These anomalies are particularly frequent during postpuberty or perimenopause. The endometrium is
thick and the stroma is dense and rich in fibroblastic cells that
deposit collagen fibers and thick fibers of reticulin. These
tubes are proliferative, but because of long-acting stimulation,
their epithelia show few mitoses and form invaginations projecting
into the lumen that seem to unwind and obliterate it completely.
In rare cases, the tubes evaginate into the stroma producing
adenomatous hyperplasias. A curettage should be done in these
cases, and ovulation should be induced in preference to treatment
with progesterone.

In associated estrogenic stimulation with normal luteal
impregnation, the endometrium is markedly hyperplastic, with
secretory glands. The dense stroma respond poorly to progesterone.

Associated moderate, and permanent estrogenic stimulation
with irregular and weak luteal influence is characterized by a
polypoid endometrium, and is known as florid glandulocystic hyperplasia. The dense, thick stroma is occupied by cells with round
nuclei and thin cytoplasm. Mitoses are absent. The tubes are
numerous, irregularly dispersed, of unequal size and most often
are found and dilated, but rarely cystic. The lining epithelium
is flattened, mitoses are rare and a slight apical secretion is
observed. The response to progesterone gradually disappears,
either because the plasma level diminishes, or because progesterone
receptors decrease at the same time that estrogenic stimulation
diminishes and estrogen receptors are not yet reconstituted.
Glandulocystic hyperplasia then progresses into an inert phase.

A high or persistent progesterone level acting on the endometrium is clinically manifested by "irregular shedding". The
decidualized endometrium exfoliates in wide, necrotic, blood
infiltrated bands. Expulsion lasts during the entire menstrual
cycle, prolonged by the persistence of fibrinolysis.

"Membranous dysmenorrhea" is recognized by evacuation of a
uterine case in one or several episodes and is often accompanied
by violent uterine contractions. Bleeding and polynuclear and
lymphocytic infiltrates are localized in the area of separating
the spongy zone. The diagnosis of "very precocious abortum" is
not always erroneous because excessive secretion of progesterone
is related either to a persistent corpus luteum (the corpus

luteum remains functional for several days, resulting in a longer
than normal cycle or to missed implantation).

DISCUSSION

    The foregoing histologic studies are borne out by various
parameters, i.e., temperature curves, vaginal smears, plasma
steroid concentrations and steroid receptors. The lowest point
on the basal body temperature curve indicates the time of ovulation.
A slow rise of temperature taking more than 2 days to reach 37°C
may be considered abnormal.

    Biphasic curves have been studied in more detail (Jolivet
and Gautray, 1978). We have distinguished one group of women with
cycles having two phases of equal duration, and in whom the temperature rise took more than 2 days to reach 37°C, indicating luteal
deficiency. In another group, the two phases were of unequal duration: a short luteal phase after a normal follicular phase; a
short luteal phase after a long follicular phase; or a normal
luteal phase after a long follicular phase causing delayed ovulation.

    Vaginal smears alone, are difficult to interpret. However,
they are particularly useful in anovulatory cycles (amenorrheas
and oligomenorrheas) since the degree of pyknosis correlates with
the level of plasma estrogens. The curve of the pyknotic index
usually remains low in estrogen deficiencies, but it may be elevated
"en plateau", sometimes with a sudden fall followed by bleeding.
In oligo- or amenorrhea, it may be of moderate, but persistent level.

    A sin curve indicates cyclic falls in estrogens without cyto-
logic evidence of luteal action and cyclic bleeding. These latter
two curves are observed in polypoid hyperplasias or in the initial
phase of the glandulocystic hyperplasia.

    In biphasic cycles, the smears, correlated with other para-
meters, confirm ovarian disturbances. In simple luteal insufficiency, an elevated pyknotic index persists during the prolonged
follicular phase, but delayed and/or insufficient cellular plicature and agglutination are evident. In luteal insufficiency
with persistent estrogenic action, the fall of the pyknotic index
is slight, and plicature and agglutination are practically absent.
In the absence of other ovulatory signs, this picture would
indicate an anovulatory cycle.

    In precocious interruption of the corpus luteum, luteal
signs (always weak) appear early and persist until day 20 of the
cycle. Afterward, two phenomena may occur, singly or simultaneously: a new rise in the pyknotic index, and/or cytolysis.

Measurement of plasma steroid concentrations during the course of the cycle enables determination of the intensity and extent of estrogen and progesterone action. In luteal insufficiency, the progesterone level is low, but it is even lower when estrogen persists and induces mitoses. In all types of pathological luteal defects, the preovulatory estrogen peak is slightly low, and the level of estradiol during the luteal phase is insufficient. Thus, the normal functioning corpus luteum must be preceded by a well-stimulated follicle. In a few perimenopausal women with histologically normal endometria, on days 24 and 25 plasma concentrations of progesterone are practically zero. It is probable that this anomaly is caused by an early interruption of the corpus luteum. In fact, early ovulation accompanied by early interruption of progesterone secretion may produce this abnormality, which has been observed after interruption of estro-progestative contraception.

The study of luteal defect involved cases of isolated luteal insufficiency with a delay of more than 2 days (compared with the normal cycle), and luteal insufficiency associated with persistent or predominant estrogen activity. The level of cytosolic and nuclear estradiol receptors is always lower than that observed during the late secretory phase of a normal cycle, which in turn is lower than that of the postovulatory phase. In cases of isolated luteal defect, the level of progesterone receptors is very low, but is nearly normal when estrogenic action persists. In both cases, the level of nuclear progesterone receptors is always lower than the normal cycle. Thus, in all cases of luteal deficiency, nuclear progesterone receptors are reduced and the level of estradiol receptors remains low because of inadequate levels of plasma estrogens. However, in cases of luteal insufficiency with persistent estrogens, the number of cytosolic progesterone receptors is increased, probably because of persistent estrogen stimulation, even though it is lower than normal.

Variations in endometrial receptor levels are thus based on the levels of the plasma steroids, estrogen and progesterone. Hence, luteal insufficiency corresponds to a complex disturbance of ovarian steroidogenesis (Bayard et al., 1978). All these parameters and the study of the relationships between steroids, plasma globulins and cellular receptors of the endometrium, lend strong histologic credibility to endometrial biopsy (Gautray et al., 1980).

CONCLUSION

The histology of the endometrium in ovarian disorders has been corroborated by the clinical picture and by plasma concentrations of endometrial steroids and hormonal receptors. Consequently, endometrial biopsy taken between days 20 and 23 of the cycle permits determination of the degree of luteal insufficiency, and, an exact view of ovarian activity.

REFERENCES

Bayard, F., Damilano, S., Robel, P., and Baulieu, E., 1978, Cytoplasmic and nuclear estradiol and progesterone receptors
in human endometrium, J. Clin. Endocr. Metab., 46:635.
Dallenbach-Hellweg, G., 1975, "Endometrium," 2nd ed., Springer-
Verlag, Berlin.
Gautray, J.P., Jolivet, A., de Brux, J., Tajchner, G., Robel, P.,
and Mouren, P., 1980, Clinical investigation of the men-
strual cycle. III. Clinical, endometrial and endocrine
aspects of luteal defect, J. Clin. Endocr. Metab. (in press).
Hachiya, S., Kusuhara, K., and Hosoda, H., 1976, The significance
of dysfunctional endometrium in sterility, in: "Recent
Advances in Human Reproduction," A. Campos da Paz, V.A.
Drill, M. Hayashi, W. Rodriguez, and A.V. Schally, eds.,
Excerpta Medica Elsevier, Amsterdam, Oxford, New York, p. 47.
Hamperl, H., 1954, Ueber die endometrialen granulozyten (endometrial
körnchenzellen), Klin. Wschr., 32:665.
Horta, J.L.H., Fernandez, J.G., De Leon, B.S., and Cortez-Gallegos,
V., 1977, Direct evidence of luteal insufficiency in women
with habitual abortion, Obstet. Gynecol., 49:705.
Jolivet, A., and Gautray, J.P., 1978, Clinical investigation of
menstrual cycle. I. Diagram of the normal menstrual cycle,
Fertil. Steril., 29:40.
Keller, D.W., Keist, W.G., Askin, F.B., Johnson, L.W., and Strickler,
R.C., 1979, Pseudocorpus luteum insufficiency: a local
defect of progesterone action on endometrial stroma, J.
Clin. Endocr. Metab., 48:127.
Noyes, R.W., Hertig, A.T., and Peck, J., 1950, Dating the endo-
metrial biopsy, Fertil. Steril., 1:3.
Parr, M.B., and Parr, E.L., 1974, Uterine luminal epithelium: pro-
trusions mediate endocytosis, not apocrine secretion in
the rat, Biol. Reprod., 11:220.
Philippe, E., Renaud, R., and Gandar, R., 1965, Le cycle endometrial normal biphasique, Rev. Fse. Gynecol., 60:405.
Richart, R.M., and Ferenczy, A., 1974, Endometrial morphologic
response to hormonal environment, Gynec. Oncol., 2:180.
Rosenfeld, D.L., and Ramon-Garcia, G., 1975, Endometrium biopsy
in the cycle of conception, Fertil. Steril., 27:1088.
Seegar-Jones, G., 1977, The clinical evaluation of ovulation and
the luteal phase, J. Reprod. Med., 18:139.
Wynn, R.M., and Woolley, R.S., 1967, Ultrastructural cycle changes
in the human endometrium. II. Normal post ovulatory phase,
Fertil. Steril., 18:721.

LUTEAL INSUFFICIENCY: ENDOMETRIAL AND ENDOCRINE CORRELATES[*]

Jean Pierre Gautray, Marie Charles Colin, Jean Paul
Vielh, and Jean de Brux

Department of Obstetrics and Gynecology
University of Paris Val de Marne
CHIC, 94010 Creteil Cédex France

Institut de Pathologie et de Cytologie Apliquée
53 rue des Belles Feuilles
75116 Paris

A critical analysis of the most important papers on luteal
insufficiency and a clinical study support this article. It is
divided into two parts: spontaneous and induced luteal insufficiency. This clinical situation is of interest because it is
characteristic of ovarian control disorders and is seldom of
ovarian origin. Moreover, infertility is a result of this abnormality and its incidence is probably greater than usually reported.

Luteal insufficiency (L.I.) is a controversial topic among
gynecologists and endocrinologists. The reason may be that it is
an ill-defined abnormality: clinical criteria are not considered
as pathognomonic or biologic patterns. The diversity of terms
used to refer to this functional disorder (luteal defect, insufficiency, inadequacy, short luteal phase) is a result of this
uncertainty. We shall use luteal insufficiency in this study.
Analysis of the literature demonstrates the slow progressive
accumulation of data on this subject; Jones initially described
the anomaly in 1949, and Strott et al. (1970) identified the short
luteal phase as the most typical, or even the exclusive, pattern
of this disorder. This firmly held opinion has had two consequences: (1) important, but less obvious patterns of this disorder
have been neglected; but (2) important endocrine data have been

---
[*]Supported by the Faculté de Medecine de Creteil

established (Jones et al., 1970; Sherman and Korenman, 1974).
More recently, this menstrual cycle abnormality has been correlated with a greater incidence of abortion (Horta et al., 1977;
Poulson and Bryner, 1977; Yip and Sung, 1977). Since 1976, hormonal data have pointed to a central endocrine origin for this
ovarian control disorder (Wilks et al., 1976).

PHYSIOPATHOLOGY OF LUTEAL INSUFFICIENCY

In this study, we shall try to correlate several parameters
to provide a better definition of L.I. Endometrial patterns have
been the basis for diagnosis of L.I. and basal body temperature
(BBT) charts and hormone measurements have been correlated to them.
The BBT chart is of clinical significance in suspected L.I.;
steroid measurements have yielded more precise information on
follicle and corpus luteum secretory activity and the physiopathology of this disorder. In addition, endometrial steroid
receptors have been used to evaluate the consequences of L.I. on
the molecular biology of the endometrium.

MATERIALS AND METHODS

Menstrual disorders and/or infertility were investigated in
a population of 328 outpatients at our institution. For this
retrospective study, 135 women were excluded for medical reasons
or incomplete records; 90 had normal cycles, 11 of which were fertile cycles; and 103 were considered to have L.I., 15 of which
were either lost to follow-up or incompletely investigated and so
were also removed from the study. Therefore, the remaining 88
patients were used as the basis of this report (26.8% of the total).
These abnormal menstrual cycles were compared to the 79 normal
cycles (90 normal minus 11 fertile cycles).

Patient age varied from 19 to 44 years, a mean of 29.5 years.
In all patients, investigation included BBT, endometrial biopsy,
and 17β-estradriol and progesterone measurements. In addition,
estrogen and progesterone receptors were quantitated in 41 endometrial samples.

BBT charts were kept for three or more cycles in 70% and for
two cycles in 30% of these patients to ascertain whether this
menstrual cycle anomaly was repetitive or even permanent. The
ovulation date was presumed to be the lowest point before the
luteal rise in temperature. The charts were studied more carefully than usual, i.e., the length of both follicular and secretory phases was estimated and plotted on a previously calculated
nomogram of a normal cycle (Jolivet and Gautray, 1978). This type
of evaluation allowed detection not only of short luteal phases,
but also of delayed ovulation. Moreover, the BBT was considered

abnormal if a slow rise requiring more than 2 days to reach or
exceed 37° C was observed.

Endometrial biopsy was performed on days 21, 22, or 23 of the
cycle for several reasons. First, it is the period of maximum
progesterone level in normal cycles. Second, it is the time when
implantation is possible (the end result of progesterone secretion),
the best date being day 21. Third, later in the cycle, even if
progesterone has allowed for maximum stromal change, the time for
implantation is over, and premenstrual tissue transformation is
evident. Histological evaluation was done according to the criteria
of Noyes et al. (1978), which apply to normal cycles. The aim was
to achieve endometrial dating at $\pm$ 1-day precision, as compared to
the patterns of a theoretical 28-day cycle with two well-balanced
phases. In cases that seemed out of phase, histologic evaluation
required estimation of the sequential and combined effects of estrogen and progesterone. These effects could be estimated not only
by the histological characteristics of endometrial cells, but also
by the characteristic development of conjunctive spines and spiral
arteries.   When characteristics typical of estrogenic influence
were pronounced during the luteal phase (persistent mitosis, abundant nuclear chromatin, poor differentiation of conjunctive spines,
loosely organized stroma), the observed anomaly was designated
persistent estrogenic influence (PEI) (see J. de Brux, Chapter 3).

The number of serum samples per patient varied. An average of
10 samples per cycle were examined from the 79 women with normal
cycles and from the 88 cases of L.I.; sera were frozen at -20° for
later use. Plasma 17β-estradiol and progesterone levels were
assayed by radioimmunoassay. The characteristics of the antisera
and the methodology of the assays are published elsewhere (Raynaud
et al., 1974; Gautray et al., 1980). LH, FSH, and prolactin were
not measured during this investigation.

The procedure for estimating steroid receptor levels in endometrial samples has already been described (Bayard et al., 1978).
Measurement of endometrial steroid receptors was possible in 41 of
our L.I. cases, which compared to identically collected specimens
of normal cycles. Calculation of the statistical significance of
means was performed by the nonparametric rank test of Wilcoxon.

RESULTS AND DISCUSSION

Some of the results of this study may be discussed with relation to generally held opinions; however, several aspects of L.I.
require a more hypothetical discussion. In a recent study Wentz
(1980) concluded that biopsy of all patients presenting with infertility appeared to be the most reasonable means of diagnosing luteal
phase defect, although the patients were not similar. She performed

210 endometrial biopsies in 149 patients and detected a 19% incidence of L.I. This figure is much higher than the 3% reported
by Jones (1976).

The incidence of L.I. was even higher in our study, but
probably for different reasons. The 328 patients we investigated
complained of menstrual disorder and/or infertility, thus, the
probability of a functional gynecologic disorder could be more
important. Such a population can be submitted to statistical
analysis. Luteal insufficiency was diagnosed in 88 patients
(26.8%) on the basis of very well-defined, and very acute histological patterns. Three patterns were significant: (1) a very
thin endometrium with rare, scattered glandular tubes, poor glycogen density, rare conjunctive spines and spiral arteries with
thin walls was observed in only 4 cases (Fig. 1); (2) a delay in
endometrial maturation of at least 2 days, with abnormal distribution of intracellular glycogen, poor or absent intraluminal secretion and persistent or predominant estrogen-influenced endometrium was the most commonly encountered pattern; and (3) exclusively

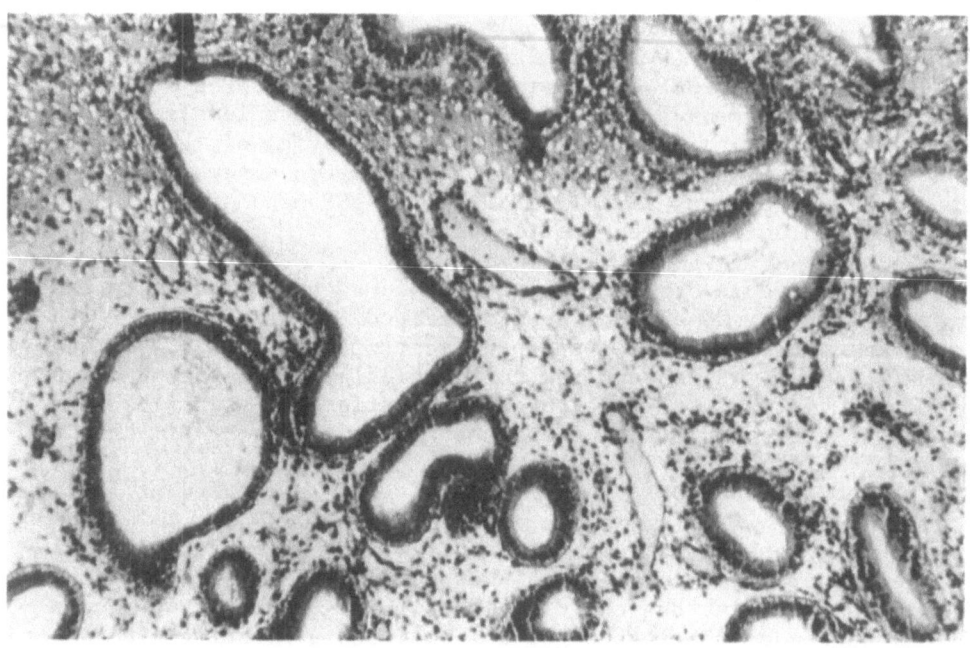

Fig. 1. Typical pattern of endometrium in short luteal phase.

in short luteal phase cases, rare, but enlarged glands that had a
small quantity of intraluminal glycogen and stromal edema were
observed. Individual cells were small, elongated or reticular and
spiral arterioles were rare and poorly differentiated (Fig. 2).

According to this very precise analysis of endometrial patterns,
pure L.I., a deficiency independent of progestational influence,
was ascertained in 41 cases (12.5%) and L.I. with persistent estrogenic influence was discerned in 47 patients (14.3%). This particular L.I. pattern has not been described before and it appears that
this dual disorder induces clinical and endocrine aspects of this
syndrome.

Although histological patterns are essential for the diagnosis
of luteal defect (Soules et al., 1977; Taubert, 1978), endometrial
biopsy is not performed routinely on patients with menstrual disorders or in the work up of infertility patients. This may be a
factor in the reported low incidence of L.I.

BBT charts were not used only to determine the length of the
cycle and the onset of ovulation. Cycles in which endometrial
biopsy suggested L.I. (88 cases) were scrutinized. The presumed

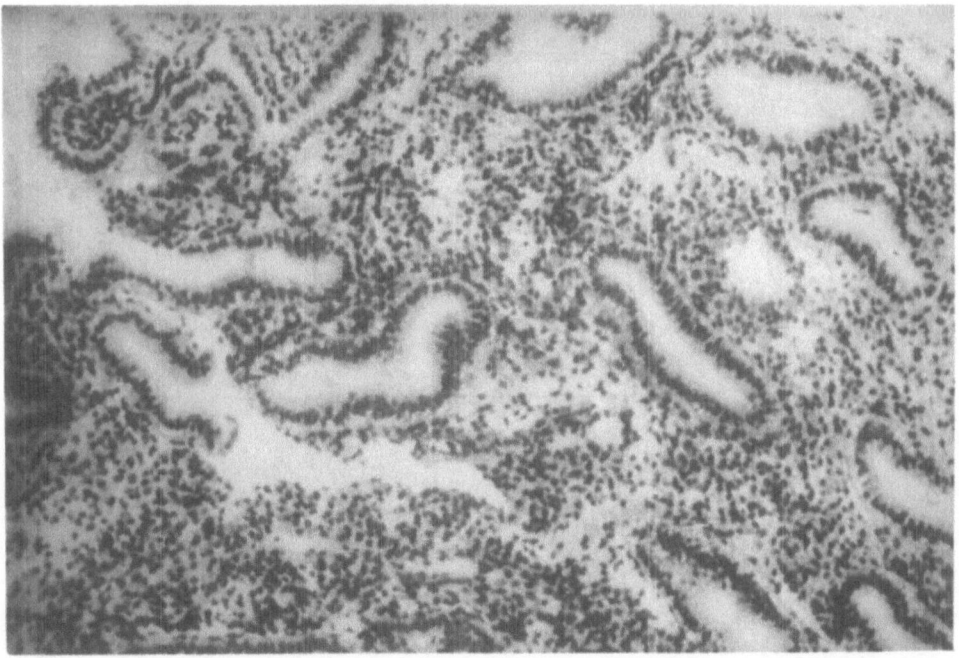

Fig. 2. Typical pattern of delayed endometrium maturation with
persistent and excessive estrogenic influence.

ovulation day was excluded for pre- and postovulatory phases
(lowest point before temperature rise to 37° C). The length of
each phase was calculated in days and then plotted on a nomogram
of the normal cycle (Jolivet and Gautray, 1978). Each cycle was
defined by a point. The tolerance limit of the population was
elliptical and contained the characteristic point of each normal
cycle; therefore, it was considered the area of normalcy (Fig. 3).

    This unusual and precise attention to BBT charts in L.I. cases
allowed us to identify two groups of patients. The first included
29 cases (8.8% of the total) that fell within the normal ellipse
of the nomogram (Fig. 4A). Although both phases appear well balanced, the length (Gautray et al., 1980) of the luteal phase was
often borderline. In all cases, a slow thermal rise reaching or
exceeding 37° C was observed which extended over 2 days, and in

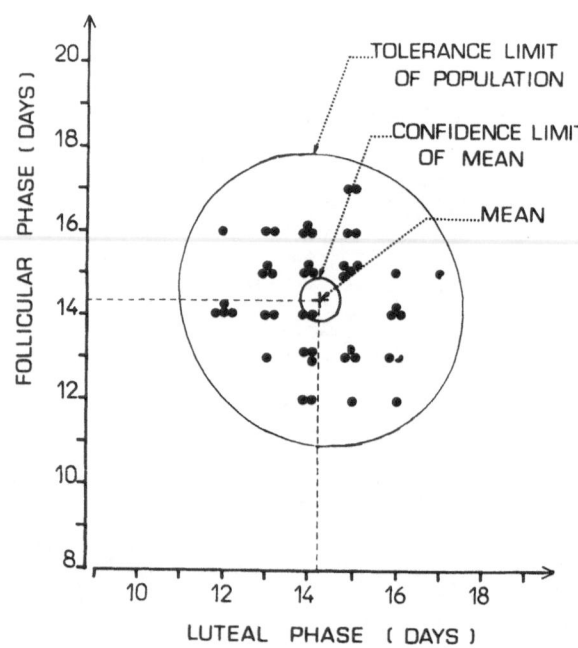

Fig. 3. Diagram of normal cycle (nomogram) calculated from both
         phase lengths (in days) of 46 clinically normal cycles
         which immediately preceded fertile cycles, in patients
         wishing pregnancy. A second species linear regression
         method was used for statistical analysis, each cycle being
         defined by 2 values (for follicular and luteal phases),
         which are all included within the two concentric ellipses
         of confidence and tolerance limits.

A   B

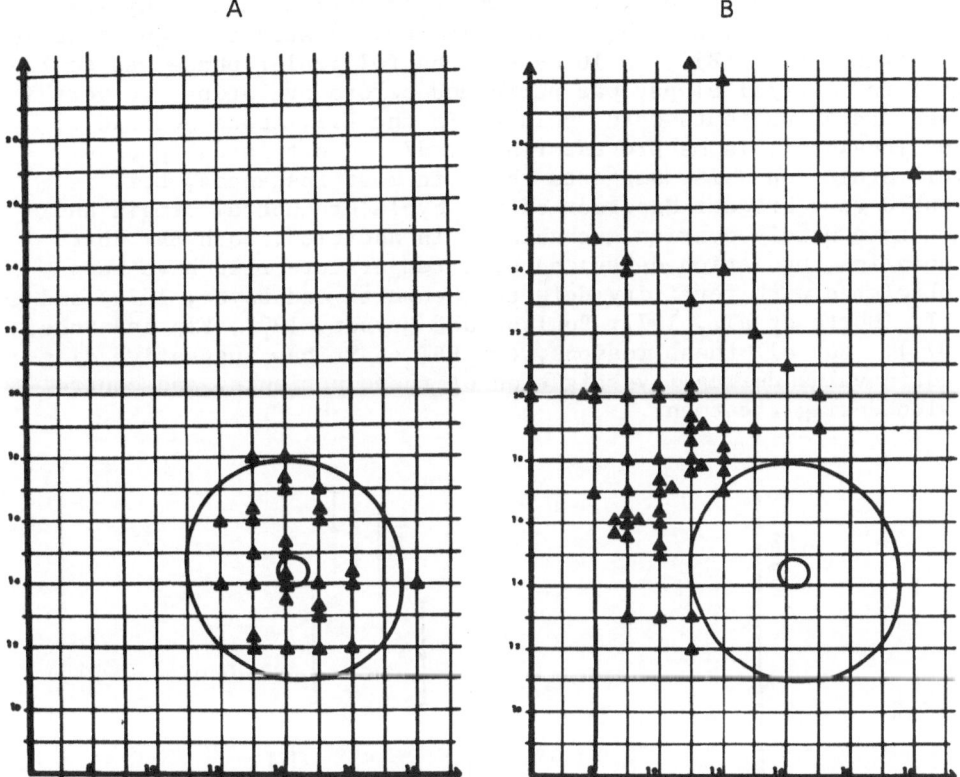

Fig. 4. BBT charts: characterization of normal cycle according to
         the length of both phases (see Fig. 3). (A) values
         observed in 29 cycles are in the ellipse of normalcy;
         (B) values of 59 other cycles are not in ellipse.

some instances lasted for 5 days. The second group included 59
cases (18% of the total): all fell outside the normal ellipse
(Fig. 4B). A wide discrepancy in the length of both phases was
obvious, i.e., short luteal phases with normal follicular phases
(20 cases, 6%), and delayed ovulation, long follicular phases with
short (19 cases, 5.8%) or normal luteal phases (20 cases, 6%).

    This statistical analysis of the cycle demonstrates that the
usual consideration of the cycle length is inadequate to appreciate
the different clinical aspects of L.I. In both patient groups the
average length of the cycles was equivalent: 28.7 days in the
first group (25 to 32 days) and 29.4 days in the second group
(23 to 45 days).

Such BBT chart analysis allowed speculation on the physiopathology of this disorder, since it demonstrated a frequent delay of
ovulation (Fig. 4B). In 10 cases, the follicular phase was so long
that endometrial biopsy was performed before ovulation. Nevertheless, these patients were included in the L.I. cases because of
low levels of plasma progesterone, and/or short luteal phases.
This ovulation delay suggests that, in most instances, L.I. is
linked to a broader disorder of the cycle of central origin and
not to a purely ovarian mechanism. In addition, both BBT chart
anomalies, ovulation delay and slow temperature rise have been
correlated with fertility defects (Schwartz and Boyce, 1974) and
L.I. (Horta et al., 1977; Poulson and Bryner, 1977; Yip and Sung,
1977). For all these reasons, the BBT chart has suggestive diagnostic value during investigation of these patients, and control
value during treatment.

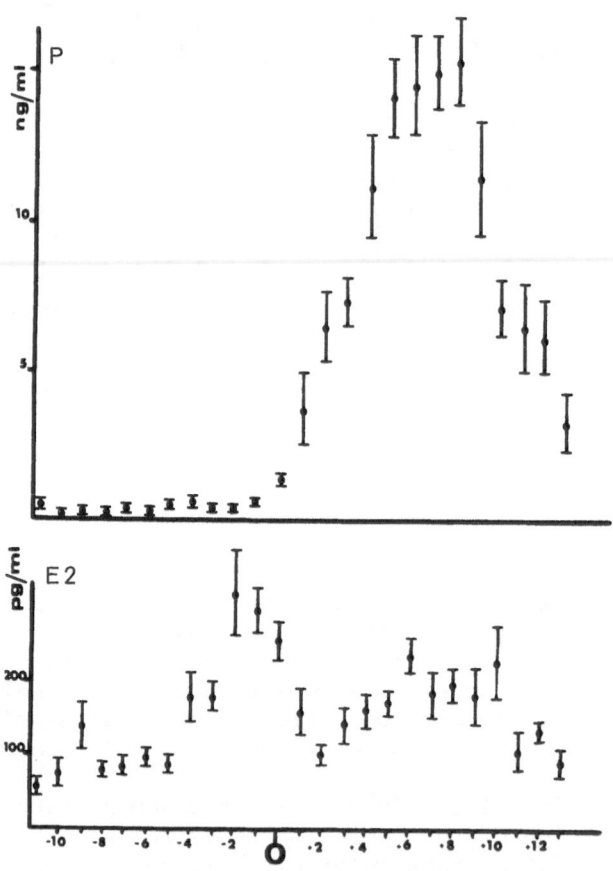

Fig. 5.   Mean values and s.e.m. during normal cycles. The cycle is
          synchronized according to the presumed date of ovulation.

Plasma estradiol and progesterone values of both L.I. groups were compared with data obtained during normal cycles. The number of patients and the number of measurements (more than 900 for each steroid) allow statistical analysis and comparison of these data. The means (± 1 s.e.m.) were calculated each day and plotted on different diagrams (Tables 1 and 2; Figs. 5 to 8).

Table 1. Plasma Progesterone Values from the Ovulation Day
(means and s.e.m.)

Progesterone (ng/ml)

Days from Ovulation	Normal			Luteal defect			Luteal defect + P.E.I.		
D 0	1.32	±	0.16	1.21	±	0.22	1.70	±	0.20
D + 1	3.72	±	1.18	3.39	±	0.78	2.89	±	0.34
D + 2	6.54	±	1.26	4.04	±	0.59	5.57	±	0.47
D + 3	7.48	±	0.85	3.98	±	0.84	6.65	±	0.95
D + 4	11.33	±	1.8	8.19	±	1.25	9.01	±	1.44
D + 5	14.34	±	1.46*	10.88	±	1.74	8.75	±	2.25*
D + 6	14.69	±	1.79*	9.45	±	1.62*	9.43	±	1.34*
D + 7	15.14	±	1.46*	10.69	±	2.21*	8.09	±	1.33*
D + 8	15.56	±	1.58*	11.89	±	1.94*	7.98	±	0.85*
D + 9	11.51	±	1.99	7.89	±	1.83	8.67	±	2.75
D + 10	7.29	±	1.03	7.39	±	1.85	8.21	±	2.42
D + 11	6.64	±	1.47	5.66			3.01	±	1.04
D + 12	6.27	±	1.16	5.85			5.58	±	1.85
D + 13	3.37	±	0.99				3.54	±	1.79

The ovulation day was postulated according to the BBT chart and progesterone rise. Statistical difference between normal and luteal defect groups is shown by *; for more precision, see Fig. 7.

Table 2. Plasma Estradiol Values from the
Ovulation Day (means and s.e.m.)

17β-Estradiol (pg/ml)

Days from Ovulation	Normal	Luteal defect	Luteal defect + P.E.I.
D – 11	55.5 + 10.65	85.5 + 44.07	66.14 + 14.62
D – 10	72.62 + 17.67	75.37 + 16.91	64.14 + 8.24
D – 9	138 + 35.96	54 + 35.47	121.40 + 41.75
D – 8	79.5 + 14.2	89.57 + 38.06	83.62 + 24.69
D – 7	83.7 + 16.51	88 + 15.82	87.5 + 9.07
D – 6	99.42 + 13.69	119.5 + 28.90	118.46 + 12.16
D – 5	86 + 13.9	*125.41 + 16.72	148.07 + 22.87
D – 4	179 + 33.41	121.5 + 19.8	196.78 + 27.69
D – 3	179 + 22.67	233.23 + 36.46	246.25 + 39.50
D – 2	316.94 + 58.64	244 + 43.53	175.86 + 28.69
D – 1	294.72 + 26.67	275.31 + 31.77	217.4 + 32.85
0	255.19 + 27.62	#292.92 + 47.18	*154.88 + 25.22
D + 1	159.36 + 36.60	#188.29 + 37.93	* 94.10 + 11.97
D + 2	105.63 + 15.10	132.92 + 22.45	*144.09 + 21.62
D + 3	144.25 + 23.9	163.16 + 36.75	134.89 + 20.25
D + 4	161.5 + 23.3	#219.2 + 39.69	125.5 + 24.19
D + 5	170.65 + 21.94	169.25 + 28.07	164.64 + 38.30
D + 6	234.35 + 27	*155.31 + 17.92	*158.26 + 24.89
D + 7	185.93 + 31.63	134.25 + 21.21	*119.83 + 9.41
D + 8	195 + 21.95	208 + 33	*129.6 + 15.26
D + 9	179 + 42	165 + 43	277 + 47.42
D + 10	233.16 + 49.9	157.4 + 26.56	178 + 53.94
D + 11	106.09 + 29.32		120.5 + 40.01
D + 12	132.66 + 14.12		158.57 + 50.04
D + 13	92.4 + 19.52		100 + 25.37
D + 14	121.5 + 47.61		74.66 + 25.97

The ovulation day was postulated according to the BBT chart and
progesterone rise. Statistical difference between normal and
luteal defect groups is shown by *; between each group of luteal
defect by #; for more precision, see Fig. 8.

Progesterone values were lower in both L.I. groups than in
normal groups and this difference was even more marked in cases
of L.I. with PEI (Figs. 5 to 8). Days 5 to 8 postovulation,
which include the period of implantation, demonstrated the most

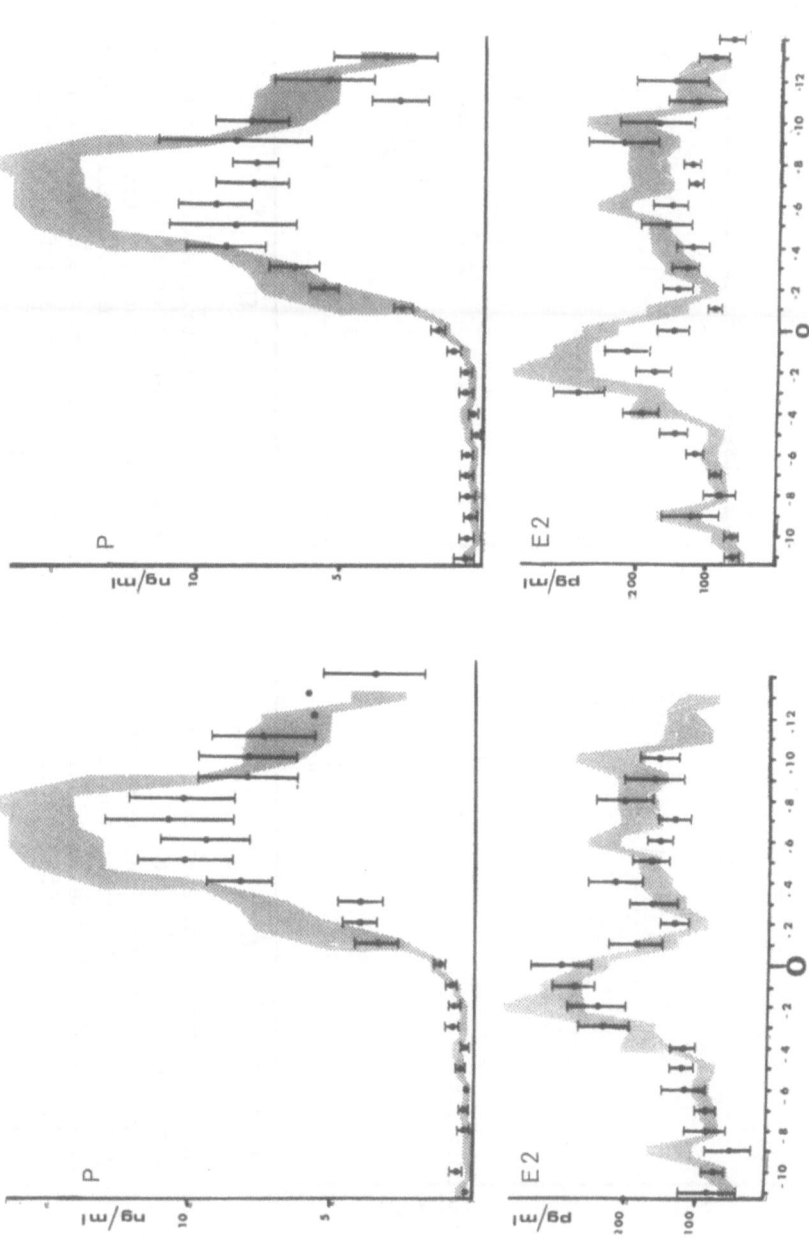

Fig. 6. Mean values and s.e.m. during cycle in L.I. The cycles are synchronized according to the presumed date of ovulation. The shaded area corresponds to the s.e.m. of these steroid concentrations during normal cycles. On the left, pure L.I.; on the right, L.I. with persistent estrogenic influence.

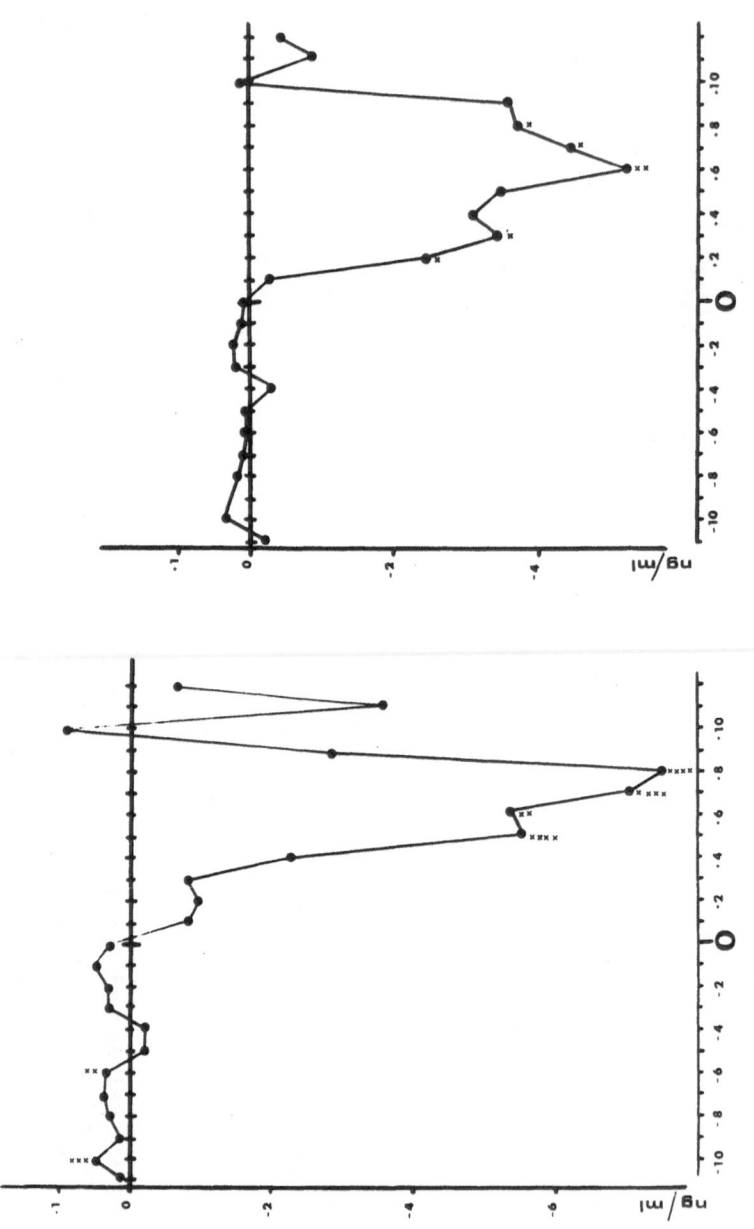

Fig. 7.   This diagram represents the difference of means of plasma progesterone levels between
normal cycles and pure L.I. (left) and L.I. with persistent estrogenic influence
(right). The zero line corresponds to normal cycles, and the positive or negative
difference demonstrates the discrepancy of values in L.I. cases. Daily mean values
were calculated, the statistical significance is plotted on the diagram and appre-
ciated from x (p<0.05)   to xxxx (p<0.001).   The progesterone defect is more severe
in cases of L.I. with P.E.I. on a histological basis.

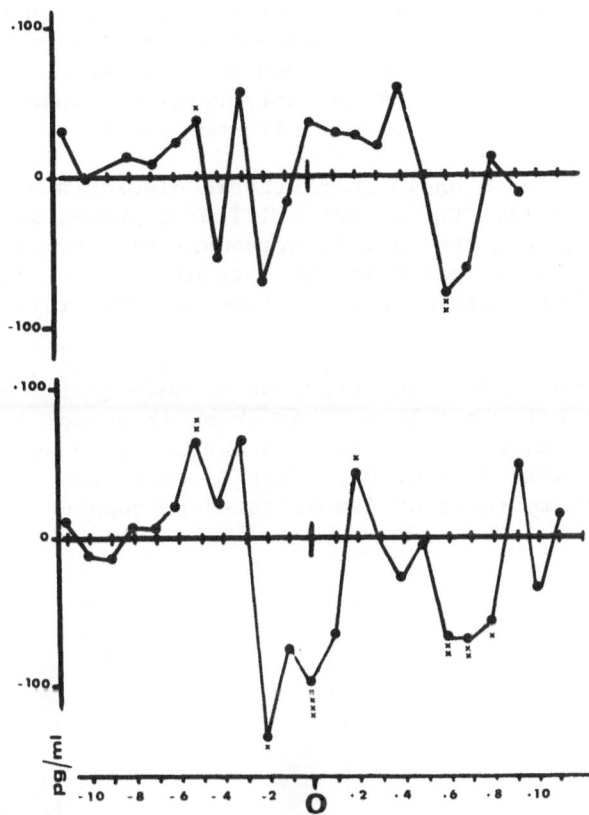

Fig. 8.   This diagram represents the difference of means of plasma
          estradiol levels between normal cycles and pure L.I.,
          (upper panel), and L.I. with persistent estrogenic influ-
          ence (lower panel). Estradiol defect is obvious during
          luteal phase of both groups during the preovulatory peak
          and more severe in cases of L.I. with P.E.I.

significant deficiency. Moreover, this statistical comparison
demonstrated low estradiol levels during the luteal phase and also
an abnormal, low preovulatory peak in both L.I. groups.

     Analysis of plasma steroid values has shown that individual
values are widely scattered, and progesterone levels may be very
different from one patient to another: high levels within the
first 3 days postovulation, followed by a rapid drop during the
next few days, low levels during a normal length luteal phase,
normal levels for a short time (short luteal phase). Patients
with histologically demonstrated L.I. cannot be placed in homo-
geneous groups according to steroid values. Statistical analysis

only can set forth the steroid secretion defect in these patients.
Thus, a single progesterone measurement may not allow diagnosis
of L.I., and serial measurements must be done, which may not be
easy in clinical practice. This controversial opinion (Rosenfeld
and Garcia, 1976) is in agreement with our study.

Analysis has also shown that steroid disorders are noticeable
throughout the cycle. During the follicular phase, we observed an
excess of plasma progesterone, in agreement with others (Glass and
Golbus, 1978), and anomalies in the estradiol peak. These facts
also suggest a pituitary/ovarian dysfunction rather than a purely
ovarian one.

Estradiol receptor concentrations of endometrial biopsies
from women with L.I. were lower than those in normal late secretory phase, the nadir of receptor concentration in nonfertile,
normal cycles (Table 3; Fig. 9). This is consistent with the
estradiol defect measured or demonstrated throughout the cycle.

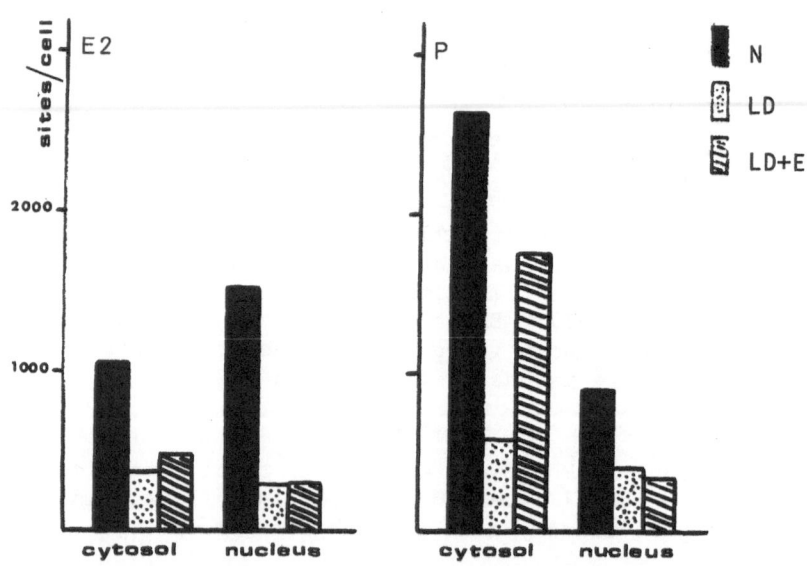

Fig. 9. Estradiol (E2) and progesterone (P) receptors in the
         endometrium, during late luteal phase of normal cycles (N),
         post ovulatory phase in pure luteal insufficiency (L.D.),
         or luteal insufficiency plus persistent estrogenic
         influence (LD+E).

Table 3. Mean Values of Estradiol and Progesterone Endometrial Receptors

	Estradiol Receptor			Progesterone Receptor		
	Total	Cytosol	Nuclei	Total	Cytosol	Nuclei
Normal cycle	2952 ± 1868	1062	1530	3541 ± 3288	2639	902
Luteal defect	674.66 ± 471.73	370.31 ± 391.30	276.87 ± 251.30	785 ± 497.87	588.24 ± 469.51	194.80 ± 205.41
Luteal defect + P.E.I.	811.53 ± 590.41	476.87 ± 339.39	300.90 ± 458.77	2149.90 ± 1984.52	1764.83 ± 1818.39	270.15 ± 517.23

Mean values of estradiol and progesterone endometrial receptors are expressed as sites/cell (one pmole per mg DNA corresponds to 4,000 sites/cell), in three different situations: (1) in normal cycle: the EB always occurs in the late secretory phase, 7-12 days postovulation; (2) in pure luteal defect; (3) in luteal defect with P.E.I.

The cytosol progesterone receptor is very low in cases of
L.I., but is close to normal in L.I. with P.E.I. This is consistent with the long, rather weak estradiol influence due to
delayed ovulation. However, in both L.I. groups, the nuclear
progesterone receptor concentration was lower than that in normal
cycles (Fig. 9; Table 3). These biological facts give new insight
into the physiopathology of L.I. The notion of low plasma estradiol levels might contradict that of P.E.I., i.e., hormone parameters would not explain histologic ones. This suggests that the
effect of estradiol on the endometrium corresponds to the estradiol level at the midfollicular phase, which is maintained (abnor-
mally) by ovulation delay and the too wide and low preovulatory
peak.

More cases have to be studied before it is possible to quan-
titate steroid receptor measurements in the diagnostic workup of
infertility patients. However, the significant discrepancies
between endometrial steroid receptor values in normal menstrual
cycles and L.I. cases demonstrate the additional importance of
endometrial biopsy.

Two aspects of L.I., its frequency and origin, deserve further discussion. The actual frequency of L.I. should be recon-
sidered, although it is difficult to establish. The 3.5% incidence reported by Jones (1976) in cases of infertility contrasts
with the 35% incidence observed in patients suffering habitual
abortion (Horta et al., 1977; Poulson and Bryner, 1977; Yip and
Sung, 1977; Glass and Golbus, 1978). The figures published recently
by Wentz (1980), 19 to 29.5%, are in agreement with our 26.8%. This
incidence figure may be due to more precise histologic interpretation and simultaneous use of different parameters in patient screen-
ing. The correlation of endometrial abnormalities with BBT anomalies and steroid hormone dysfunction in our study, supports this
concept of L.I., and, an even higher frequency is possible.

Different clinical or biological aspects of L.I. suggest a
central endocrine or even neuroendocrine disorder. Gonadotropin
levels could not be measured in this study because of the difficulty in synchronizing the time of blood sampling for outpatients
scheduled for clinic visits. However, perturbations in gonadotropin secretion have been demonstrated in luteal phase defects
in rhesus monkeys (Wilks et al., 1976; Nass et al., 1979), have
been suspected in humans (Sobonale et al., 1978), and more recently, have been demonstrated in a few clinical cases (Aksel, 1980).
To date, no statistical study has been performed.

Prolactin was not measured in our study, and there has been
controversy over the level of prolactin during the entire cycle
or during the periovulatory period, and its influence on luteal

function (Corenblum et al., 1976; Seppala et al., 1976; Aksel, 1980). In a shorter and more recent study of L.I., we observed both situations: L.I. with and without functional hyperprolactinemia (unpublished data).

Hypophysial dysfunction is most probable in recurrent L.I. The benefit provided by neuroendocrine but not by hormonal therapy (Gautray et al., 1978) represents additional indirect evidence for neuroendocrine involvement in these cases. This study suggests a greater frequency of L.I. than generally recognized and emphasizes the necessity of accurate histologic investigation of the endometrium. Occasional L.I. is probably borderline to the normal physiology of the menstrual cycle (Aksel, 1980). In contrast, recurrent L.I. is a cause of infertility and miscarriage. The inadequacy of the endometrium may explain both the poor fertility of these patients and the higher abortion rate because proper implantation is impossible.

INDUCED LUTEAL INSUFFICIENCY

Progestogens have been considered as substitutes for progesterone after initial investigations in rodents. Their action in primates and humans is quite different, and Johannson (1971) demonstrated the luteolytic influence of norethisterone. This fact is not widely known, and prescription of progestogen during the second part of the cycle to treat different functional disorders is not uncommon. For substitutive therapy, only progesterone itself should be prescribed (Jones, 1976; Soules et al., 1977; Taubert, 1978; Wentz, 1980), or a modified molecule of progesterone, i.e., retroprogesterone.

A different, and more recent aspect of induced L.I. comes from the so-called paradoxical effect of luteinizing hormone-releasing hormone (LHRH). It has been demonstrated that high doses of LHRH are able to induce abortion or infertility in rats (Corbin and Beattie, 1976). Later, testicular atrophy was induced by the same method (Auclair et al., 1977; Pelletier et al., 1978). Similar results were obtained by high doses of LH or hCG. Hormonal investigation of gonad endocrine cells demonstrated a desensitization of these cells a few hours after the induced abnormal LH peak (Labrie et al., 1979). Similar results have been shown in monkeys and humans when LHRH is administered within 72 hours after ovulation (Raynaud et al., 1978). A drop in progesterone concentration is first observed and is associated with a shortened luteal phase. The phenomenon is regularly induced, but its amplitude and efficiency cannot be foreseen (Fig. 10).

This methodology has been proposed as a contraceptive method, using either repeated high doses of synthetic LHRH after the LH surge (250 µg every 4 hr x 5 on one or two consecutive days;

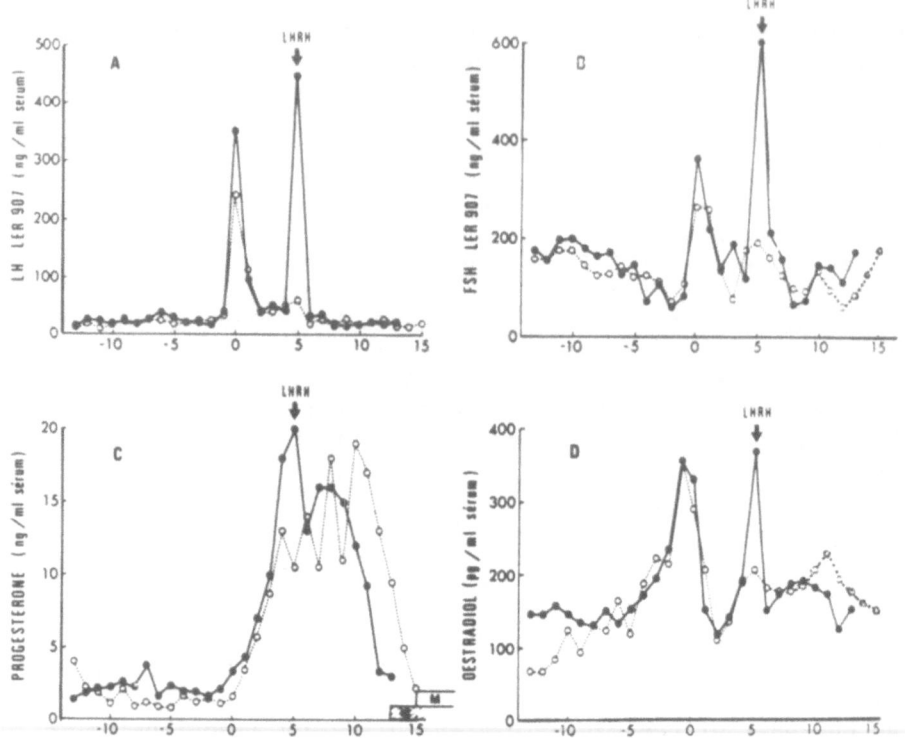

Fig. 10.   Effect of 5 subcutaneous 250 µg doses of LHRH given
           every 4 h on day 7 after the LH surge of serum LH (A),
           FSH (B), progesterone (C), and 17β-estradiol (D), and
           at time of menses in normal women (Raynaud et al., 1978).

Raynaud et al., 1978), or more potent analogs (D-Ala[6], Des-Gly-
NH$_2$[10]) LHRH ethylamide, or D-Ser (TBU)[6]LHRH 1-9 ethylamide.
Before implementation of clinical trials with such agonists in
humans, studies in nonhuman primates, investigating efficacy, side
effects and reversibility should be more conclusive (Wickings et
al., 1980).   In monkeys and human males, results appear rather
disappointing (Berquist et al., 1979; Wickings et al., 1980).   No
statistical evaluation is possible at the present time.   However,
at least two questions have been raised by clinical experimentation.
The first concerns the contraceptive efficacy of such a method.
The second is the safety of a method that induces L.I. with which
mastopathies are known to be frequently associated (Sitruk-Ware
et al., 1979).   Finally, there is disagreement whether metrorrhagia
or pelvic pain may be induced by these practices.

In conclusion, it appears that L.I. is more frequent than usually estimated; spontaneous L.I. is the consequence, at the gonadal level, of a hypophyseal and/or hypothalamic perturbation of ovarian control, whereas induced L.I. is an interesting example of desensitization of granulosa cells. L.I. may be a chance event, and then, only a physiologic circumstance, but when it is repetitive or permanent, it undoubtedly causes infertility and abortion. In our opinion, this menstrual cycle abnormality requires more attention.

REFERENCES

Aksel, S., 1980, Sporadic and recurrent luteal phase defects in cyclic women: comparison with normal cycles, Fertil. Steril., 33:372.

Auclair, C., Kelly, P.A., Coy, D.H., Schally, A.V., and Labrie, F., 1977, Potent inhibitory activity of [D-leu, des-Gly-NH$_2^{10}$] LHRH ethylamide on LH/hCG and PRL testicular receptor levels in the rat, Endocrinology, 101:1890.

Bayard, F., Damilano, S., Robel, P., and Baulieu, E.E., 1978, Cytoplasmic and nuclear estradiol and progesterone receptors in human endometrium, J. Clin. Endocrinol. Metab., 46:635.

Berquist, G., Nillius, S.J., Bergh, T., Skarin, G., and Wide, L., 1979, Inhibitory effects on gonadotropin secretion and gonadal function in men during chronic treatment with a potent stimulatory LHRH analogue, Acta Endocrinol. (Kbh), 91:601.

Corbin, A., and Beattie, C.W., 1976, Regulation of female mammalian fertility with luteinizing hormone releasing hormone (LHRH) and related analogues, in: "Ovulation in the Human," P.G. Crosignani and D.R. Mishell, ed., Academic Press, New York, p. 193.

Corenblum, B., Pairaudeau, N., and Shewshuk, A.B., 1976, Prolactin hypersecretion and short luteal phase defects, Obstet. Gynecol., 4:486.

Gautray, J.P., de Brux, J., Tajchner, G., Robel, P., and Mouren, M., 1981, Clinical investigation of the menstrual cycle. III. Clinical, endometrial, and endocrine aspects of luteal defect, Fertil., Steril., in press.

Gautray, J.P., Jolivet, A., Goldenberg, F., Tajchner, G., and Eberhard, A., 1978, Clinical investigation of the menstrual cycle. II. Neuroendocrine investigation and therapy of the inadequate luteal phase, Fertil. Steril., 29:275.

Glass, R.H., and Golbus, M.T., 1978, Habitual abortion, Fertil. Steril., 29:257.

Horta, J.L.H., Fernandez, J.G., De Leon, B., and Cortes-Gallegos, V., 1977, Direct evidence of luteal insufficiency in women with habitual abortion, Obstet. Gynecol., 49:705.

Johannson, E.D.B., 1971, Depression of the progesterone levels in women treated with synthetic gestagens after ovulation, Acta Endocrinol. (Kbh), 68:779.

Jolivet, A., and Gautray, J.P., 1978, Clinical investigation of the menstrual cycle. I. Diagram of the normal menstrual cycle, Fertil. Steril., 29:40.

Jones, G.E.S., 1949, Some newer aspects of management of infertility, J. Am. Med. Assoc., 141:1123.

Jones, G.S., 1976, The luteal phase defect, Fertil. Steril., 27:35.

Jones, G.E.S., and Madrigal-Castro, V., 1970, Hormonal findings in association with abnormal corpus luteum function in the human: the luteal phase defect, Fertil. Steril., 21:1.

Labrie, F., Auclair, C., Cusan, L., Lemay, A., Belanger, A., Kelly, P.A., Ferland, L., Azadian-Boulanger, G., and Raynaud, J.P., 1979, Inhibitory effects of treatment with LHRH or its agonists on ovarian receptor levels and function, in: "Ovarian Follicular and Corpus Luteum Function," Plenum Press, New New York.

Nass, T.E., Dierschke, D.J., Clerk, J.R., Meller, P.A., and Schillo, K.K., 1979, Luteal phase deficiencies in peripubertal rhesus monkeys: mechanistic considerations, Adv. Exp. Med. Biol., 112:519.

Noyes, R.W., Hertig, A., and Rock, J., 1950, Dating the endometrial biopsy, Fertil. Steril., 1:3.

Pelletier, G., Cusan, L., Auclair, C., Désy, L., and Labrie, F., 1978, Inhibition of spermatogenesis in the rat by treatment with [D-Ala^6, des-Gly-NH$_2$^{10}]LHRH ethylamide, Endocrinology, 103:641.

Poulson, A.M., and Bryner, W.A., 1977, The obstetric complication of the infertility patient, Obstet. Gynecol., 49:174.

Raynaud, J.P., Azadian-Boulanger, G., and Bucourt, R., 1974, Anticorps spécifiques de l'estradiol, J. Pharmacol. (Paris), 5:27.

Raynaud, J.P., Azadian-Boulanger, G., Mary I., Mouren, M., Lemay, A., Ferland, L., Auclair, C., and Labrie, F., 1978, Action luteolytique de LHRH chez la ratte, la guenon et la femme, in: "L'implantation de l'oeuf," F. Du Mesnil Du Buisson, A. Psychoyos, and K. Thomas, ed., Masson et Cie, Paris, p. 273.

Rosenfeld, D.L., and Garcia, C.R., 1976, A comparison of endometrial histology with simultaneous plasma progesterone determination in infertile women, Fertil. Steril., 27:1256.

Schwartz, D., and Boyce, A., 1974, Avortement et montée thermique du cycle menstrual. Résultat d'une étude épidémiologique, in: "Les Accidents Chromosomiques, et la Reproduction," 1 vol INSERM, Paris, p. 341.

Seppala, M., Hirvonen, E., and Rauta, T., 1976, Hyperprolactinemia and luteal insufficiency, Lancet, 1:229.

Sherman, B.M., and Korenman, S.G., 1974, Measurement of serum LH, FSH, estradiol and progesterone in disorders of the human menstrual cycle: the inadequate luteal phase, J. Clin. Endocrinol. Metab., 39:145.

Sobonale, O., Lenton, E.A., Francis, B., Cooke, I.D., 1978, Comparison of plasma steroid and gonadotropin profiles in spontaneous cycles in which conception did and did not occur, Br. J. Obstet. Gynaecol., 85:460.

Soules, M.R., Wiebe, R.H., Aksel, S., and Hammond, C.B., 1977, The diagnosis and therapy of luteal phase deficiency, Fertil. Steril., 28:1033.

Strott, C.-A., Cargille, C.M., Ross, G.T., and Lipsett, M.B., 1970, The short luteal phase, J. Clin. Endocrinol. Metab., 30:246.

Taubert, H.D., 1978, Luteal phase insufficiency, in: "Female Infertility," Karger, Basel, p. 78.

Wentz, A.C., 1980, Endometrial biopsy in the evaluation of infertility, Fertil. Steril., 33:121.

Wickings, E.J., Zaidi, P., and Nieschlag, E., 1980, Do LHRH superagonists provide an approach to male infertility control: Preclinical trials in rhesus monkeys, Acta Endocrinol. [Suppl.] (Kbh), 234:78.

Wilks, J.W., Hodgen, G.D., and Ross, G.T., 1976, Luteal phase defects in the rhesus monkey: the significance of serum FSH: LH ratios, J. Clin. Endocrinol. Metab., 43:1261.

Yip, S.K., and Sung, M.L., 1977, Plasma progesterone in women with a history of recurrent early abortions, Fertil. Steril., 28:151.

FUNCTIONAL ASPECTS OF ENDOMETRIAL HYPERPLASIA AND CARCINOMA

Rodrigue Mortel[*], Paul Robel, and Etienne Emile Baulieu

Unité de Recherches sur le Métabolisme Moléculaire
et la Physio-Pathologie des Stéroides
L'Institut National de la Sante et de la Recherche
Médicale
(U33 INSERM) and ER 125 CNRS
78 rue du General Leclerc
94270 Bicetre, France

[*]Present Address:
Milton S. Hershey Medical Center
The Pennsylvania State University
Department of Obstetrics and Gynecology
Division of Gynecologic Oncology
Hershey, Pennsylvania

INTRODUCTION

Hyperplasia and adenocarcinoma of the endometrium occur primarily in women subjected to long-standing, unopposed estrogenic stimulation (MacMahon, 1974; Smith et al., 1975; Gusberg, 1976). In addition, obesity, diabetes, hypertension and late menopause are encountered in a high percentage of these patients, suggesting that endometrial carcinoma may be an endocrinopathy (Kelley, 1973). Consequently, in the past several years, gynecologists, endocrinologists, and biochemists have developed a considerable interest in the study of hyperplastic and malignant endometria. Studies have been designed to establish the hormonal aspects of neoplastic endometrium and its functional relationship to normal endometrium.

The endometrium is a target organ for sex steroid hormones. These hormones interact with target cells to effect cellular growth and differentiation through specific intracellular receptors. This interaction leads to differential transcription, translation and

the resultant biologic response (Jensen and DeSombre, 1972;
Baulieu, 1975). Indeed, in a few experimental models, the magnitude of hormonal effects has been positively correlated with
receptor concentration. Likewise, no hormonal effect has been
elicited in cells lacking measurable receptors.

Measurement of estradiol and progesterone receptors in the
cytosol and nuclei of normal endometrial samples revealed that
total cellular concentration of both receptors increases during
the early proliferative phase and reaches a peak in the late proliferative phase of the menstrual cycle. Subsequently, by the
end of the cycle, the receptor concentration of both hormones
decreases to levels lower than those observed in the early proliferative phase (Levy et al., 1980). In addition, when the data
on receptor measurement were compared with serum levels of estradiol and progesterone in the same patients, the receptor concentration of both hormones correlated positively with serum estradiol during the proliferative phase. However, as serum progesterone
rises during the secretory phase, there is a progressive decline
in progesterone receptor concentration (Levy et al., 1980). This
paper reviews the functional relationship between normal, hyperplastic, and malignant endometrium and evaluates their hormone
dependence by receptor concentration, enzyme activity, prostaglandin measurement and clinical data.

Endometrial Hyperplasia

Clinical significance. Endometrial hyperplasia includes a
wide range of epithelial abnormalities characterized by varying
degrees of morphologic alterations and cellular changes (Dallenbach-Hellweg, 1971; Gore, 1973; Welch and Scully, 1977). A variety
of terms have been applied to this histological picture. However,
depending upon morphologic changes and primarily cytologic alterations, these lesions have been classified as cystic hyperplasia,
adenomatous hyperplasia, atypical endometrial hyperplasia or
carcinoma in situ (Welch and Scully, 1977).

From an endocrinological standpoint, patients who develop
endometrial hyperplasia are identical to those commonly affected
by endometrial cancer. Hyperplasia is most prevalent in women of
perimenopausal age when anovulatory cycles are frequent (Sherman
et al., 1976; Welch and Scully, 1977). It is also observed in
young women with anovulatory cycles, polycystic ovarian syndrome
or estrogen-producing ovarian tumors (Kistner, 1979). Numerous
epidemiologic, experimental and clinical studies have implicated
unopposed estrogen with the development of endometrial hyperplasia
and its progression to carcinoma (Hertig and Sommers, 1949; Beutler
et al., 1963). A variety of retrospective and prospective studies
indicate that in 85% of these women, the hyperplasia remains the

same or will not recur after uterine curettage (Beutler et al., 1963; Kistner, 1979). However, the histologic picture progresses to endometrial carcinoma in 5 to 15% of adenomatous hyperplasia cases and 1 to 2% of cystic hyperplasia cases (Gusberg, 1947; Richardson and MacLaughlin, 1978a).

Receptors in anovulatory cycles. Since anovulatory cycles constitute a common denominator in patients with endometrial hyperplasia, estradiol and progesterone receptors in both cytosol and nuclei were measured in a group of patients with this aberration in the menstrual cycle (Levy et al., 1980). The technique utilized was reported in detail by Bayard et al. (1978) and allows measurement of total occupied and unoccupied receptor sites. The high concentrations of estradiol and progesterone receptors found were comparable to those observed in the late proliferative phase of the cycle. In addition, most receptors were concentrated in the nuclear fraction, hence the low cytoplasmic to nuclear ratio of estradiol receptor found in these endometria. The concentration of nuclear progesterone receptor, however, was significantly lower than in preovulatory endometrium. The increase in nuclear estrogen and cytoplasmic progesterone receptors in endometria of anovulatory patients may explain at a biochemical level the hyperestrogenic state and the risk of hyperplasia and carcinoma in this patient group (Levy et al., 1980).

Receptors in hyperplasia. Based on data accumulated on sex steroid receptor regulation in normal endometrium, high levels of estrogen and progesterone receptors should be expected in endometrial hyperplasia. Indeed, Jänne, et al. (1978) confirmed the presence of high levels of cytoplasmic estrogen and progesterone receptors in hyperplastic endometria. Likewise, Gurpide et al. (1976) measured estradiol receptors in endometrial hyperplasia and observed concentrations similar to those in proliferative phase endometria or in women treated with exogenous estrogen. Crocker et al. (1974) also demonstrated an increase in estrogen receptor in endometrial hyperplasia; values reached those observed during the late proliferative phase. It is known that a characteristic effect of estradiol is an increase in progesterone receptor concentration (PRC) in the uterus (Milgrom et al., 1973). Therefore, PRC measurement may be utilized as a biochemical indicator of the response of tumor cells to estrogen. Indeed, when the PRC:estrogen nuclear receptor ratio was taken as an index of cellular sensitivity to estrogenic stimuli, a gradual increase of the index was demonstrated from proliferative phase to postmenopausal endometria, and a further increase was seen in endometrial hyperplasia. Interestingly, the index was highest in adenomatous hyperplasia (Richardson and MacLaughlin, 1978b).

Estradiol 17β-hydroxysteroid dehydrogenase (17β-HSD) in endometrial hyperplasia. This enzyme is present in the endometrium and is responsible for converting estradiol to estrone. Its activity is quite low during the proliferative phase of the cycle, but increases in the presence of progesterone (Tseng and Gurpide, 1975). In vivo experiments (Pollow et al., 1975a; Tseng and Gurpide, 1975) have demonstrated increased enzymatic activity when progesterone or progestagens are administered to women during the proliferative phase of the menstrual cycle. Since 17β-HSD activity is mediated by progesterone or progestins, and since plasma progesterone is undetectable in patients with endometrial hyperplasia, a low enzyme activity would be expected in hyperplastic endometrium. Indeed, Gurpide et al. (1976) reported that activity values in hyperplastic endometrium were similar to those found in proliferative endometrium.

Prostaglandins in endometrial hyperplasia. In 1977, Mortel et al. first reported on prostaglandin levels in the tumor bed of gynecologic malignancies. Prostaglandins E and F (PGE and PGF) were measured in blood taken from hypogastric arteries and veins of women with cervical and endometrial cancers by radioimmunoassay. No appreciable differences were noted in the prostaglandin levels of arterial or venous blood, except in cases of endometrial cancer, where an increase in the venous blood was noted. It was thought, therefore, that endometrial cancers could synthesize prostaglandin.

Prostaglandins have been found in normal endometrium. The levels increase during the secretory phase of the menstrual cycle and reach a peak in menstrual endometrium (Demers et al., 1975). When normal and hyperplastic endometrial explants were maintained in vitro, high levels of PGE were found in the culture medium. PGE levels were higher in secretory endometrium, as previously demonstrated, and were identical in proliferative and hyperplastic endometria. Interestingly, when the same endometrial samples were maintained in culture for 2 days in the presence of 0.5 and 1.0 g/ml of progesterone, there was a significant decrease in PGE production in hyperplastic compared with proliferative or secretory endometrium (unpublished data).

Response of endometrial hyperplasia to progestagens. From a clinical standpoint, many investigators have studied the response of hyperplastic endometria to hormonal treatment and have reported endometrial sensitivity to progestagens (Kistner, 1959, 1961; Kjorstad, 1977; Richardson and MacLaughlin, 1978c). Therefore, induction of ovulation or administration of progestagen has been recommended in the treatment of patients with various types of endometrial hyperplasia (Kistner, 1979). Kjorstad (1977) has noted, however, that a small percentage of hyperplasias do not regress and some endometria return to the original hyperplastic state when progestin treatment is discontinued. This difference in endometrial

hyperplasia response and behavior provides grounds for the application of different therapeutic treatments of these lesions. In
fact, hysterectomy is the treatment of choice in peri- or postmenopausal women with endometrial hyperplasia (particularly the
adenomatous type), whereas induction of ovulation or administration
of progesterone has been recommended for young women with diagnosed
endometrial hyperplasia.

Carcinoma of the Endometrium

Receptors in endometrial carcinoma. Progestational agents
achieve objective remission in 30 to 35% of patients with advanced
or metastatic endometrial cancer (Anderson, 1965; Kelley, 1965;
Kennedy, 1968; Reifenstein, 1971). However, based on clinical
criteria alone, it has not been possible to select those patients
likely to benefit from hormonal treatment.

The measurement of steroid hormone receptor levels in tumor
tissue has been shown to be of clinical value in managing patients
with carcinoma of the breast (McGuire et al., 1977; Allegra et al.,
1979). Tumors with high receptor levels are more likely to respond
to hormonal therapy than tumors with low receptor levels. Clinical
application of steroid receptor measurements in breast cancer has
provided much of the impetus for study of steroid receptor concentrations in endometrial cancer.

Sex steroid receptors measured in the endometrium have been
found to be under hormonal regulation (Crocker et al., 1974; Levy
et al., 1980). Estradiol increases progesterone and its receptors,
whereas progesterone has a negative effect on both estradiol and
progesterone receptor concentrations. Numerous investigators have
examined estradiol and progesterone receptor concentrations in
human endometrial carcinoma (Tseng and Gurpide, 1972; Crocker et
al., 1974; Evans et al., 1974; Pollow et al., 1975b; Young et al.,
1976; MacLaughlin and Richardson, 1978); however, these studies
are barely comparable because of methodological differences and
assay variations. Our group systematically measured the cytoplasmic
and nuclear concentrations of estradiol and progesterone receptors
in malignant endometrium to study the hormonal regulation of these
receptors and to evaluate hormone responsiveness of endometrial
cancers (Mortel et al., 1981). The results of our study indicate
that nearly all tumors contain estradiol and progesterone receptors,
but at variable concentrations. No receptors were detected in
either cytosol or nuclei of one tumor examined for estradiol or
five assayed for progesterone.

The total estradiol receptor concentration of the tumors
examined was similar to that of late proliferative phase normal
endometrium and is in agreement with findings of Tseng et al. (1977).

Crocker et al. (1974) and Pollow et al. (1975b) found estradiol
receptors in all endometrial cancer samples, whereas Grilli et al.
(1977) and Muechler et al. (1975) reported estrogen receptors in
67 to 92% of the specimens assayed. Therefore, it appears that
estradiol receptor is present in the large majority of endometrial
adenocarcinomas.

The values for progesterone receptor concentration were
relatively low and comparable to those found in the secretory
phase of normal menstrual cycles when the receptor concentration
is lowest. Consequently, the estradiol:progesterone receptor ratio
in postmenopausal endometrial carcinoma was higher than in normal
perimenopausal endometrium at any time during the menstrual cycle.

When present, most estradiol and progesterone receptors were
found in the cytosol. This finding is not surprising, considering
the low levels of circulating estrogen and progesterone in these
women. However, in six estradiol and two progesterone samples,
the receptor concentration was much higher in the nuclear fraction.
In addition, all measurable receptors were located in the nuclei
of five estradiol and two progesterone samples. Therefore, simultaneous measurement of both cytoplasmic and nuclear receptor sites
appears necessary to accurately report the total intracellular
receptor concentration in endometrial cancer.

Receptor levels and tumor grading. The relationship between
receptor levels and the grade of tumor differentiation has been
the subject of conflicting reports (Evans et al., 1974; Pollow
et al., 1975b; Gustafsson et al., 1977; Feil et al., 1979). Although
no statistical difference was observed within histologic groups,
our study and others (Tseng and Gurpide, 1972; Pollow et al., 1975b;
Young et al., 1976; Gustafsson et al., 1977; Feil et al., 1979)
have shown that well-differentiated tumors are more likely to
contain high levels of both estrogen and progesterone receptors.
Our data disagree with those of Pollow et al. (1977), who reported
a progressive increase in estradiol receptor concentration from
well- to poorly-differentiated tumors.

Sex steroid receptor regulation in endometrial carcinoma.
When serum levels of estrone, estradiol, and progesterone were
measured in postmenopausal patients with endometrial carcinoma,
the serum estradiol values were lower than the estrone values.
Serum progesterone levels were as low as those in normal postmenopausal women and were identical within histologically defined
groups. Estradiol values did not change with the degree of tumor
differentiation, but a highly significant difference (p<0.01) in
mean estrone levels of patients with well- to less-differentiated
cancers was observed. This finding supports the concept that
estrone is the main circulating estrogen in postmenopausal women
at risk of developing endometrial carcinoma. The finding that

patients with less-differentiated tumors have higher levels of estrone than those with well-differentiated cancer is interesting, although no definite explanation can be provided at this time.   It is not known to what extent the nature of the main circulating estrogen might affect morphology and differentiation of tumor cells. As formulated by Siiteri et al. (1974), there is a strong possibility that estrone plays an important role in the development of post-menopausal endometrial carcinoma.

When linear regression correlations were calculated between serum hormone levels and total estradiol and progesterone receptor concentration, there was no correlation between estradiol and progesterone receptor concentration, there was no correlation between estradiol receptor and serum estrone, estradiol, or pro-gesterone levels.   Similarly, no correlation was found between progesterone receptor and serum levels of estrone, estradiol, or progesterone.   However, in well-differentiated tumors, a positive correlation at the limit of statistical significance was observed between serum estradiol and progesterone receptor concentration. It appears, therefore, that in endometrial carcinoma, as in normal endometrium, the concentration of progesterone receptor is regulated by the level of circulating estrogens.

Biochemical evaluation of hormone dependence in endometrial carcinoma.   Several authors (Ehrlich et al., 1978; Creasman et al., 1980) have deemed progesterone receptor concentration impor-tant in assessing hormone responsiveness of endometrial carcinoma. However, studies (Evans et al., 1974; MacLaughlin and Richardson, 1978; Mortel et al., 1981) have revealed the presence of measurable amounts of estradiol and wide-ranging concentrations of progesterone receptors in most cancers.   The degree of tumor differentiation and circulating hormone levels may account for such variability.   In addition, carcinoma of the endometrium may coexist with prolifer-ative, secretory, or hyperplastic endometria.   Since benign endo-metria contain high levels of receptors, the heterogeneity of the sample assayed may make it exceedingly difficult to rule out the participation of receptor-positive nonmalignant endometrium, in spite of histologic examination performed in adjacent fragments of the sample.   Therefore, it is unlikely that an isolated measure-ment of estradiol and progesterone receptors will provide accurate assessment of hormone responsiveness in endometrial carcinoma   A dynamic test might be more helpful in establishing the hormone dependency of malignant endometrium.

Hormonal test for endometrial carcinoma responsiveness.
Gurpide and Tseng (1978) and Pollow et al. (1978) propose a hormonal test based on the measurement of 17β-HSD before and after medroxy-progesterone administration to patients with endometrial carcinoma. They report up to a fourfold increase in enzyme activity and a reduction in estrogen receptor level in tumors responsive to

progestins. In addition, Pollow et al. (1978) report that enzyme
activity values were highest in well-differentiated tumors and that
these levels decreased as the tumors became more anaplastic. These
findings suggest that measurement of 17β-HSD before and after progestin treatment in concert with determination of sex steroid
receptor concentrations may provide a reliable means to select
those patients most likely to respond to hormone therapy. However,
the question remains whether there is a means to increase the progesterone receptor concentration in malignant endometrium or to
convert progesterone receptor-negative tumors into tumors with
measurable progesterone receptors

    Because of the high estrogen:progesterone receptor ratio
found in endometrial carcinomas, and because estrogens have been
implicated in the etiology of the disease, Mortel et al. (1981)
have measured estradiol and progesterone receptors in endometrial
cancers before and after administration of the antiestrogen, tamoxifen. They demonstrated that at a dose of 40 mg/d for 5 to 7
days, tamoxifen increased the progesterone receptor concentration
in most endometrial cancer samples. Interestingly, at the same
dosage, a substantial elevation in progesterone receptor was
noted in all endometrial tumors that had very low or immeasurable
receptor concentrations before tamoxifen treatment. These findings
confirm progesterone receptor induction by tamoxifen in human
endometrium, an effect previously demonstrated in animal studies,
and suggest that such a dynamic test can predict hormonal responsiveness more accurately than a single measurement of sex steroid
receptors. In addition, receptor-negative tumors can be converted
to tumors with higher progesterone receptor levels. These results
also indicate that tamoxifen may be of therapeutic benefit to
patients with advanced or metastatic adenocarcinoma of the endometrium.

    17β-HSD in endometrial carcinoma. The activity of this
enzyme in carcinoma of the endometrium is identical to that observed
in proliferative and hyperplastic endometria (Pollow et al., 1975a;
Tseng and Gurpide, 1975), and in some studies (Pollow et al., 1975a;
Mortel et al., 1981), there is a tendency for enzymatic activity to
be higher in well-differentiated samples. Gurpide and Tseng (1978)
and Pollow et al. (1978) also measured enzymatic activity in endometrial cancer samples after administration of progestagens, and
demonstrated an increase in 17β-HSD in these tumors. The response
was significant in well-differentiated tumors; however, no change
was observed in anaplastic cancers. When the same experiments
were conducted with tamoxifen (Mortel et al., 1981) no change in
the enzyme activity was observed.

    Prostaglandin in endometrial cancer. The idea of using plasma
prostaglandin determinations as a marker in endometrial cancer
resulted from preliminary studies performed by Mortel et al. (1977).

Because some patients with endometrial carcinoma showed a significant difference in prostaglandin levels in arterial versus venous
blood samples in the tumor bed, and endometrial explants have been
found to synthesize prostaglandin when maintained in culture, we
measured levels of PGE and 6-keto-PGF$_1\alpha$, a metabolite of PGF$_2$
in blood samples taken from hypogastric arteries and veins of
patients with endometrial carcinoma of various histologic gradings.
Preliminary results indicate an increase in PGE in the
venous blood draining these tumors (Fig. 1). Likewise, the 6-keto-
PGF$_1\alpha$ levels (Fig. 2) were markedly increased in the hypogastric
vein of all patients with endometrial cancer regardless of the
histologic differentiation of the individual tumor. In addition,
when prostaglandins were measured in hypogastric arteries and veins
of patients given previous radiation therapy (Fig. 3), there was
no significant increase in prostaglandins in venous samples and
the levels of PGE in these patients were markedly lower than in
patients not given radiation treatments. The few patients in the
irradiated group that demonstrated an increase in PGE in the venous
sample had residual tumor in the uterus post irradiation. These
findings suggest that prostaglandins are synthesized by endometrial
cancers and decrease considerably with tumor destruction. Therefore,
it appears that prostaglandins may potentially serve as a tumor
marker in endometrial carcinoma.

Hormone response of endometrial carcinoma. Since the original observation by Kelley and Baker (1960) that progesterone
might be effective in the treatment of metastatic endometrial
adenocarcinoma, progestins have been routinely used in this patient
group. It is now well established that an objective regression
or arrest of tumor growth will occur in approximately one-third of
patients treated with progestins (Kennedy, 1968; Reifenstein, 1971;
Kohorn, 1976). The response rate is higher in patients with well-
differentiated tumors: precisely those tumors with the highest
progesterone receptor concentration and potential for induction of
17β- HSD activity. In anaplastic tumors, the progesterone receptor
concentration is low, the response to progestins is minimal, and
the induction of 17β-HSD activity is not possible.

One drawback of progesterone therapy in the treatment of
endometrial carcinoma is a decrease in tumor progesterone receptor
concentration after 4 weeks of treatment (Jänne et al., 1978).
Since tamoxifen can rescue progesterone receptor, it has the potential, when combined with progestins, to increase the number of
responders or the magnitude and/or duration of response in patients
with advanced or metastatic endometrial carcinoma.

CONCLUSION

From a functional standpoint, there is little difference
between hyperplastic and normal endometria. Hyperplastic endo-

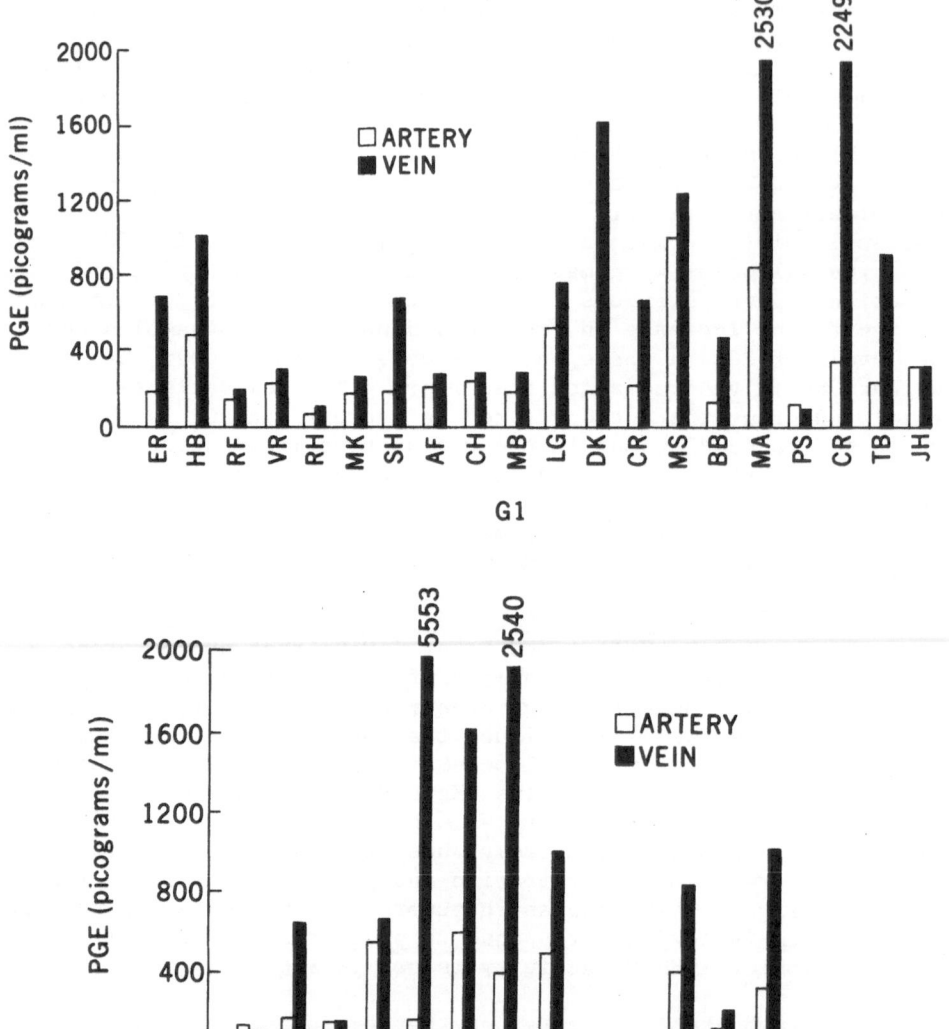

Fig. 1. PGE levels in hypogastric artery and vein across endometrial adenocarcinoma bed. G1, G2, G3 = grade 1, grade
2, grade 3.

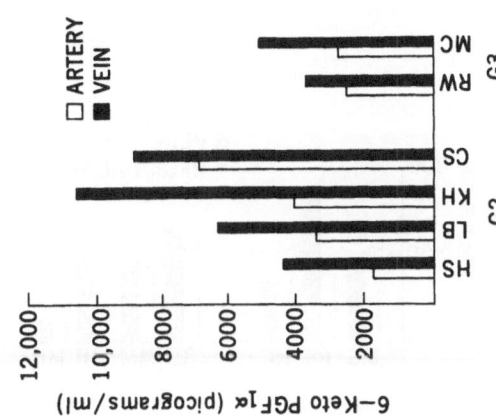

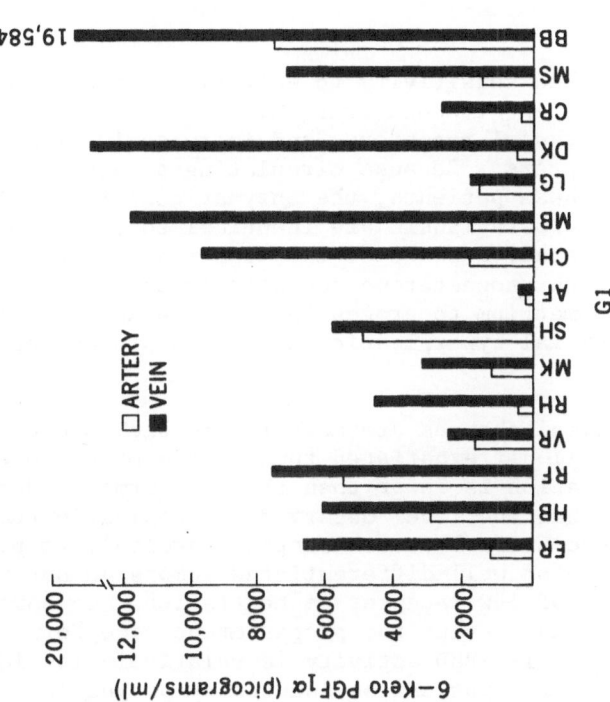

Fig. 2. 6-Keto PGF1α levels in hypogastric artery and vein across endometrial adenocarcinoma bed. G1, G2, G3 = Grade 1, Grade 2, Grade 3.

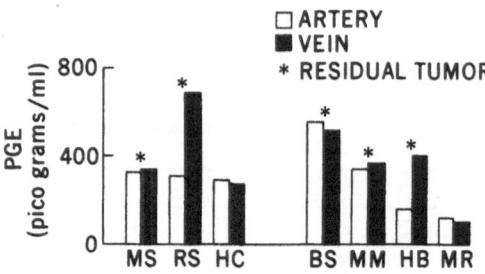

Fig. 3. PGE levels in hypogastric artery and vein across endo-
         metrial adenocarcinoma bed previously irradiated.

metrium has a marked sensitivity to estradiol, as demonstrated by
total estradiol receptor concentration, prominent nuclear distribution of the estradiol receptors, and increase in cytoplasmic
progesterone receptors. Because circulating progesterone levels
are very low in these patients, the enzymatic activity of 17β-HSD
and the level of prostaglandin are identical to those observed
during the proliferative phase of the menstrual cycle. However,
because of the high progesterone receptor content, the response of
hyperplastic endometrium to progestins is remarkable, although some
return to the initial hyperplastic state when progestins are discontinued.

    The biochemical changes are more remarkable in endometrial
carcinoma. In well-differentiated tumors, the mean progesterone
receptor concentration is lower than that of normal endometrium,
but much higher than the level observed in anaplastic tumors.
Accordingly, these Grade 1 tumors respond favorably to progestins.
A small percentage of well-differentiated tumors do not respond to
progestin because of low receptor concentration. In some of these
tumors, tamoxifen may rescue the progesterone receptor concentration. In addition, 17β-HSD activity is relatively low in well-
differentiated tumors, but can be induced by progestins. The prostaglandin level is markedly reduced in the medium when endometrial
carcinoma explants are maintained in culture in the presence of
progestins.

Anaplastic tumors contain very low concentrations of progesterone receptors and their response to progestin is minimal (Ehrlich et al., 1978). Tamoxifen does not induce progesterone receptors and 17β-HSD activity cannot be induced by progesterone. In other words, the malignant endometrium loses its functional capability as tumors progress from well- to less-differentiated types.

REFERENCES

Allegra, J.C., Lippman, M.E., Thompson, E.B., Simon, R., Barlock, A., Green, L., Huff, K.K., Do, H.M.T., Aitken, S.C., and Warren, R., 1979, Relationship between the progesterone, androgen and glucocorticoid receptor and response rate to endocrine therapy in metastatic breast cancer, Cancer Res., 39:1973.

Anderson, D.G., 1965, Management of advanced endometrial adenocarcinoma with medroxyprogesterone acetate, Am. J. Obstet. Gynecol., 92:87.

Baulieu, E.E., 1975, Steroid receptors and hormone receptivity. New approaches in pharmacology and therapeutics, J. Am. Med. Assoc., 234:404.

Bayard, F., Damilano, S., Robel, P., and Baulieu, E.E., 1978, Cytoplasmic and nuclear estradiol and progesterone receptors in human endometrium, J. Clin. Endocrinol. Metab., 46:635.

Beutler, H.K., Dockerty, M.B., and Randall, L.M., 1963, Precancerous lesions of the endometrium, Am. J. Obstet. Gynecol., 86:433.

Creasman, W.T., McCarty, K.S., Sr., Barton, T.K., and McCarty, K.S., Jr., 1980, Clinical correlates of estrogen- and progesterone-binding proteins in human endometrial adenocarcinoma, Obstet. Gynecol., 55:363.

Crocker, S.G., Milton, P.J.D, and King, R.J.B., 1974, Uptake of [6,7-^{3}H]oestradiol-17β by normal and abnormal human endometrium, J. Endocrinol., 62:145.

Dallenbach-Hellweg, G., 1971, "Histopathology of the Endometrium," Springer-Verlag, New York, p. 77.

Demers, L.M., Halbert, D.R., Jones, D.E.D., and Fontana, J., 1975, Prostaglandin F levels in endometrial jet wash specimens during the normal human menstrual cycle, Prostaglandins, 10:1057.

Evans, L.H., Martin, J.D., and Hahnel, R., 1974, Estrogen receptor concentration in normal and pathological human uterine tissues, J. Clin. Endocrinol. Metab., 38:23.

Ehrlich, C.E., Cleary, R.E., and Young, P.C.M., 1978, The use of progesterone receptors in the management of recurrent endometrial cancer, in: "Endometrial Cancer," M.G. Brush, R.B.T. King, R.W. Taylor, ed., Baillière-Tindal, London p. 258.

Feil, P.D., Mann, W.J., Mortel, R., and Bardin, C.W., 1979, Nuclear progestin receptor in normal and malignant human endometrium, J. Clin. Endocrinol. Metab., 48:327.

Gore, H., 1973, Hyperplasia of endometrium, in: "The Uterus," H.J. Norris, A.T. Hertig, and M.R. Abell, ed., Williams and Wilkins, Baltimore, p. 255.

Grilli, S., Ferrari, A.M., Gola, G., Rochetta, R., Orlandi, C., and Prodi, G., 1977, Cytoplasmic receptors for 17β-estradiol, 5-dihydrotestosterone and progesterone in normal and abnormal human uterine tissue, Cancer Lett., 2:247.

Gurpide, E., Gusberg, S.B., and Tseng, L., 1976, Estradiol binding and metabolism in human endometrial hyperplasia and adeno-carcinoma, J. Steroid Biochem., 7:891.

Gurpide, E., and Tseng, L., 1978, Potentially useful tests for responsiveness of endometrial cancer to progestogen therapy, in: "Endometrial Cancer," M.G. Brush, R.B.T. King, and R.W. Taylor, ed., Báilliere-Tindal, London, p. 52.

Gusberg, S.B., 1947, Precursors of corpus carcinoma estrogens and adenomatous hyperplasia, Am. J. Obstet. Gynecol., 54:905.

Gusberg, S.B., 1976, The individual at high risk for endometrial cancer, Am. J. Obstet. Gynecol., 126:535.

Gustafsson, J.A., Einhorn, N., Elfstrom, G., Nordenskjold, B., and Brange, O., 1977, Progestin receptor in endometrial carcinoma. Progesterone receptor in normal neoplastic tissues, Prog. Cancer Res. Ther. 4:299.

Hertig, A.T., and Sommers, S.G., 1949, Genesis of endometrial carcinoma; study of prior biopsies, Cancer, 2:946.

Jänne, O., Kauppila, A., Kontula, K., Syrjälä, P., Vierikko, P., and Vihko, R., 1980, Female sex steroid receptors in human endometrial hyperplasia and carcinoma, in: "Steroid Receptors and Hormone Dependent Neoplasia," J.L. Wittliff, and O. Dapont, ed., Masson, p. 37.

Jensen, E.V., and DeSombre, E.R., 1972, Mechanism of action of the female sex hormones, Annu. Rev. Biochem., 41:203.

Kelley, R.M., and Baker, W.H., 1960, Progestational agents in the treatment of carcinoma of the endometrium, N. Engl. J. Med., 264:216.

Kelley, R.M., and Baker, W.H., 1965, The role of progesterone in human endometrial cancer, Cancer Res., 25:1190.

Kelley, R., 1973, Progestins, in: "Cancer Medicine," J. Holland and E. Frei, III, ed., Lea and Febiger, Philadelphia, p. 923.

Kennedy, B.J., 1968, Progestogens in the treatment of carcinoma of the endometrium, Surg. Gynecol. Obstet., 127:103.

Kistner, R.W., 1959, Histological effects of progestins of hyper-plasia and carcinoma in situ of the endometrium, Cancer, 12:1106.

Kistner, R.W., 1961, Hyperplasia and carcinoma in situ of the endometrium, in: "Progesterone," Brook Lodge Press, Augusta, Michigan, p. 165.

Kistner, R.W., 1979, "Gynecology: Principles and Practice," 3rd
    ed., Yearbook Medical Publishers, Chicago, p. 245.

Kjorstad, K.E., 1977, Progestagens as only treatment in premalig-
    nant changes of the endometrium. International Meeting on
    Endometrial Cancer and Related Subjects, St. Thomas, London.

Kohorn, E.R., 1976, Gestagens and endometrial cancer, Gynecol.
    Oncol., 4:389.

Levy, C., Robel, P., Gautray, J.P., de Brux, J., Verma, U., Descomps,
    B., and Baulieu, E.E., 1980, Estradiol and progesterone
    receptors in human endometrium, normal and abnormal menstrual
    cycles and early pregnancy, Am. J. Obstet. Gynecol., 136:646.

McGuire, W.L., Horwitz, K.B., Pearson, O.H., and Segaloff, A., 1977,
    Current status of estrogen and progesterone receptors in breast
    cancer, Cancer, 39:2934.

MacLaughlin, D.T., and Richardson, G., 1978, Progesterone binding
    by normal and abnormal human endometrium, J. Clin. Endocrinol.
    Metab., 42:667.

MacMahon, B., 1974, Risk factors for endometrial carcinoma, Gynecol.
    Oncol., 2:122.

Milgrom, E., Thi M.L., and Baulieu, E.E., 1973, Control mechanisms
    of steroid hormone receptors in the reproductive tract, Kar-
    olinska Symposia on Research Methods in Reproductive Endo-
    crinology, May 23-23, p. 380.

Mortel, R., Allegra, J.C., Demers, L.M., Harvey, H.A., Trautlein,
    J., Nahhas, W., White, D., Gillin, M.A., and Lipton, A.,
    1977, Plasma prostaglandins across the tumor bed of patients
    with gynecologic malignancy, Cancer, 39:2201.

Mortel, R., Levy, C., Wolff, J.P., Nicolas, J.-C., Robel, P., and
    Baulieu, E.E., 1981, Female sex steroid receptors in post-
    menopausal endometrial carcinoma and biochemical response to
    an antiestrogen, Cancer Res., in press.

Muechler, E.K., Flinckinger, G.L., Mangan, C.E., and Mikhail, G.,
    1975, Estradiol binding by human endometrial tissue, Gynecol.
    Oncol., 3:244.

Pollow, K., Boquoi, E., Lübbert, H., and Pollow, B., 1975a, Effect of
    gestagen therapy upon 17β-hydroxysteroid dehydrogenase in
    human endometrial adenocarcinoma, J. Endocrinol., 67:131.

Pollow, K., Lübbert, H., Boquoi, E., Kreuzer, G., and Pollow, B.,
    1975b, Characterization and comparison of receptors of 17β-
    estradiol and progesterone in human proliferative endometrium
    and endometrial carcinoma, Endocrinology, 96:319.

Pollow, K., Schmidt-Gollwitzer, M., and Nevinny-Stickel, J., 1977,
    Progesterone receptors in normal human endometrium and endo-
    metrial carcinoma. Progesterone receptors in normal and neo-
    plastic tissues, Prog. Cancer Res. Ther., 4:313.

Pollow, K., Schmidt-Gollwitzer, M., Boquoi, E., and Pollow, B.,
    1978, Influence of estrogens and gestagens on 17β-hydroxy-
    steroid dehydrogenase in human endometrium and endometrial
    carcinoma, J. Mol. Med., 3:81.

Reifenstein, E.C., 1971, Hydroxyprogesterone caproate therapy in advanced endometrial cancer, Cancer, 27:485.

Richardson, G.S., and MacLaughlin, D.T., 1978a, Hormonal Biology of Endometrial Cancer, Vol. 42, UICC Technical Report Series, Geneva, p. 50.

Richardson, G.S., and MacLaughlin, D.T., 1978b, Hormonal Biology of Endometrial Cancer, Vol. 42, UICC Technical Report Series, Geneva, p. 152.

Richardson, G.S., and MacLaughlin, D.T., 1978c, Hormonal Biology of Endometrial Cancer, Vol. 42, UICC Technical Report Series, Geneva, p. 156.

Sherman, B.M., West, J.H., and Korenman, S.G., 1976, The menopause transition: analysis of LH, FSH, estradiol and progesterone concentrations during menstrual cycles in older women, J. Clin. Endocrinol., 42:629.

Siiteri, P.K., Schwarz, B., and MacDonald, P.C., 1974, Estrogen receptor and the estrone hypothesis in relation to endometrial and breast cancer, Gynecol. Oncol., 2:228.

Smith, D.C., Prentice, R., Thompson, D.J., and Herrman, W.L., 1975, Association of exogenous estrogen and endometrial carcinoma, N. Engl. J. Med., 293:1164.

Tseng, L., and Gurpide, E., 1972, Nuclear concentration of estradiol in superfused slices of human endometrium, Am. J. Obstet. Gynecol., 114:995.

Tseng, L., and Gurpide, E., 1975, Induction of human endometrial estradiol dehydrogenase by progestins, Endocrinology, 97:825.

Tseng, L., Gusberg, S., and Gurpide, E., 1977, Estradiol receptor and 17β-dehydrogenase in normal and abnormal human endometrium, Ann. N.Y. Acad. Sci., 286:190.

Welch, W.R., and Scully, R.E., 1977, Precancerous lesions of the endometrium, Hum. Pathol., 8:503.

Young, P.C.M., Ehrlich, E., and Cleary, R.E., 1976, Progesterone binding in human endometrial carcinoma, Am. J. Obstet. Gynecol., 125:353.

CONTRIBUTORS

Etienne Emile Baulieu
Lab Hormones
Bicetre, France

Catherine Blacker
Fondation de Recherche en Hormonologie
Paris, France

Viviane Casimiri
Laboratoire Physiologie de la Reproduction
Bicetre, France

Marie Charles Colin
Department of Obstetrics and Gynecology
University of Paris Val de Marne
Creteil Cedex, France

Jean de Brux
Fondation d'Enseignement et de Recherche en Histo et Cytopathologie
Paris, France

Jean Pierre Gautray
Department of Obstetrics and Gynecology
University of Paris Val de Marne
Creteil Cedex, France

Catherine Malet
Laboratoire Physiologie de la Reproduction
Bicetre, France

Dominique Martel
Laboratoire de Physiologie de la Reproduction
Bicetre, France

Rodrigue Mortel
M.S. Hershey Medical Center
Pennsylvania State University
Hershey, Pennsylvania   USA

Khalil Nahoul
Fondation de Recherche en Hormonologie
Paris, France

Alexandre Psychoyos
Laboratoire Physiologie de la Reproduction
Bicetre, France

Paul Robel
Lab Hormones
Bicetre, France

Robert Scholler
Fondation de Recherche en Hormonologie
Paris, France

Jean Paul Vielh
Department of Obstetrics and Gynecology
University of Paris Val de Marne
Creteil Cedex, France